CAMBRIDGE LIBRARY COLLECTION

Books of enduring scholarly value

Earth Sciences

In the nineteenth century, geology emerged as a distinct academic discipline. It pointed the way towards the theory of evolution, as scientists including Gideon Mantell, Adam Sedgwick, Charles Lyell and Roderick Murchison began to use the evidence of minerals, rock formations and fossils to demonstrate that the earth was older by millions of years than the conventional, Bible-based wisdom had supposed. They argued convincingly that the climate, flora and fauna of the distant past could be deduced from geological evidence. Volcanic activity, the formation of mountains, and the action of glaciers and rivers, tides and ocean currents also became better understood. This series includes landmark publications by pioneers of the modern earth sciences, who advanced the scientific understanding of our planet and the processes by which it is constantly re-shaped.

The Climate of London

The 'student of clouds' Luke Howard (1772–1864) published this work of statistics on weather conditions in London in two volumes, in 1818 and 1820. Howard was by profession an industrial chemist, but his great interest in meteorology led to his studies on clouds (also reissued in this series), and his devising of the system of Latin cloud names which was adopted internationally and is still in use. Volume 2 contains a preface in which Howard discusses the reasons for the order of presentation of his material, which, he disarmingly admits with hindsight, might have been improved. The tabular material in this volume, supplied with notes and commentary citing published reports from around Europe, comes from observations made at Tottenham in the period 1817–19. This historic material will be of interest to environmental scientists as well as to those interested in the history of meteorology.

Cambridge University Press has long been a pioneer in the reissuing of out-of-print titles from its own backlist, producing digital reprints of books that are still sought after by scholars and students but could not be reprinted economically using traditional technology. The Cambridge Library Collection extends this activity to a wider range of books which are still of importance to researchers and professionals, either for the source material they contain, or as landmarks in the history of their academic discipline.

Drawing from the world-renowned collections in the Cambridge University Library and other partner libraries, and guided by the advice of experts in each subject area, Cambridge University Press is using state-of-the-art scanning machines in its own Printing House to capture the content of each book selected for inclusion. The files are processed to give a consistently clear, crisp image, and the books finished to the high quality standard for which the Press is recognised around the world. The latest print-on-demand technology ensures that the books will remain available indefinitely, and that orders for single or multiple copies can quickly be supplied.

The Cambridge Library Collection brings back to life books of enduring scholarly value (including out-of-copyright works originally issued by other publishers) across a wide range of disciplines in the humanities and social sciences and in science and technology.

The Climate of London

Deduced from Meteorological Observations

VOLUME 2

LUKE HOWARD

CAMBRIDGE UNIVERSITY PRESS

Cambridge, New York, Melbourne, Madrid, Cape Town,
Singapore, São Paolo, Delhi, Mexico City

Published in the United States of America by Cambridge University Press, New York

www.cambridge.org
Information on this title: www.cambridge.org/9781108049528

© in this compilation Cambridge University Press 2012

This edition first published 1818
This digitally printed version 2012

ISBN 978-1-108-04952-8 Paperback

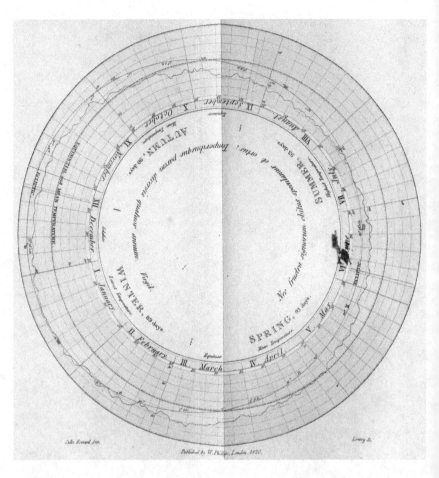

The material originally positioned here is too large for reproduction in this reissue. A PDF can be downloaded from the web address given on page iv of this book, by clicking on 'Resources Available'.

CLIMATE OF LONDON,

DEDUCED FROM

𝕸𝖊𝖙𝖊𝖔𝖗𝖔𝖑𝖔𝖌𝖎𝖈𝖆𝖑 𝕺𝖇𝖘𝖊𝖗𝖛𝖆𝖙𝖎𝖔𝖓𝖘,

MADE AT DIFFERENT PLACES

IN THE

NEIGHBOURHOOD OF THE METROPOLIS.

———

By LUKE HOWARD.

———

IN TWO VOLUMES.

———

VOL. II.

Containing (besides a Preface to the second volume) the remainder of the Series of Observations, up to Midsummer 1819 : an Account of the Climate, under the heads of *Temperature*, Barometrical *Pressure*, *Winds*, *Evaporation*, *Moisture* by the Hygrometer, *Rain*, *Lunar periods :* with a *Summary* of Results, in the order of the seasons ; *General Tables ;* and a copious *Index.*

———

LONDON:

PRINTED AND SOLD BY W. PHILLIPS, GEORGE YARD, LOMBARD STREET :
SOLD ALSO BY J. AND A. ARCH, CORNHILL ; BALDWIN, CRADOCK,
AND JOY, AND W. BENT, PATERNOSTER ROW ; AND J. HATCHARD,
PICCADILLY.

———

1820.

PREFACE TO THE SECOND VOLUME.

—

THE Map of my subject being at length delineated, the reader has it in his power to survey it; which he may do either in detail, in the several divisions of *Temperature*, &c. which follow, or first as a whole, in the *Summary*, where it is treated in the order of the months and seasons. On turning over the work now that it is about to be completed at press, I am sensible of some imperfections in the arrangement; which might have been made more easy for the reader, and the text less interrupted by results in figures, had the whole been reserved, till now, unprinted. The few points of theory which I have introduced here and there, might likewise have been embodied in a preliminary dissertation; the want of which will be scarcely supplied, to some readers, by the summary above mentioned. My principal apology must lie, in the want of a good model, for a design so nearly novel in character: to which may be added a strong inducement to print the several parts, as they were digested, for the sake of easy reference. In attempting to reduce to some sort of method the great mass of observations before me, I was not seldom in the case of the traveller in a South American forest, who is obliged, even where others have gone before him, to cut his way at every step through a

a 2

tangled thicket. If the *vista* be in any degree thus opened, those who may follow will scarcely grudge the labour of smoothing asperities, filling up chasms, and making plain the road to the science. With regard to mathematical discussions, with which it would have been an easy task to some, to have interspersed the work, I think it right to avow, that a limited education in that branch of science has left me unqualified to furnish them : and possibly, to men capable of applying them to the test of sound theory, the simple *data* derived from observation may prove as acceptable, as a splendid series of ready-made *demonstrations*. One thing the reader may rely on—that much care has been exercised in the plain calculations which were continually required to bring out my results. It may be proper also to remark, that for the convenience of those who may incline to take up the subject only in parts, the *Index* has been made copious and minute, to a degree which on any other consideration would have been quite superfluous.

The result of my experience is, on the whole, unfavourable to the opinion of a permanent change having taken place of later time, either for the better or the worse, in the Climate of this country. Our recollection of the weather, even at the distance of a few years, being very imperfect, we are apt to suppose that the seasons are not what they formerly were; while in fact, they are only going through a series of changes, such as we may have heretofore already witnessed, and forgotten. That the shorter periods of annual variation in the mean temperature, depth of rain, and other phenomena of the year, which will be found exhibited in this volume, may be only component parts of a larger cycle is, however, very possible. Otherwise, considering that the changes consequent on the clearing of woods, culture and drainage, with some other less obvious effects of an increased population, have probably by this time contributed their utmost to its improvement, I should venture to suppose, that our Climate is likely to remain for ages what it now is; and further that, in its great leading features, it differs little from what it was, when the present elevation of these islands above the sea was first established.

Having despatched the few remarks of this kind that were left for a preface, I may now claim the indulgence of the scientific reader for some thoughts of a more important nature. In the introduction to my earliest published observations (in 1807) I find the following remarks on the *end* and *object* of such enquiries. " Every correct register of the weather may be considered as intended for two purposes: first, as a daily record of the phenomena regarded as passing occurrences; secondly, as a continued notation of facts interesting to the philosopher, and from which he may deduce results, for the purpose of extending our knowledge of the œconomy of the seasons. This application of the subject it is desirable to encourage: for it cannot be doubted, that from views less limited we should draw conclusions less partial as to these changes, and instead of that scene of confusion, that domain of chance, which as commonly seen they present, we should discover a chain of causes and effects, demonstrative like the rest of creation, of the infinite wisdom and goodness of its Author." Athenæum, vol. i. p. 80. I should indeed regret the many hours of leisure, which I have since bestowed on this pursuit, could I not persuade myself, that these anticipations are likely to be in some measure fulfilled: that Meteorology will, by future observers, at least, be rescued from empirical mysteriousness, and the reproach of perpetual uncertainty; and will contribute its share to the support of a proposition, so well illustrated by some of the brightest names in science, that the "Almighty hand, that made the world of matter without form, hath ordered all things in measure and number and weight." Wisd. xi. 17, 20. Or, (to use more modern terms) that the Creator has, even in the course of the winds and the variations of the atmosphere, so adapted the means to the end, that amidst perpetual fluctuations, and occasional tremendous perturbations, the balance of the great machine is preserved, and its parts still move in harmony: each returning season verifying the assurance given to mankind after the deluge, "While the earth remaineth, seed-time and harvest, and cold and heat, and summer and winter, and day and night shall not cease." Gen. viii. 22.

I have occasionally observed with regret, in the writings of men of science, the continuance of a phraseology which I would gladly see exploded; which is unmeaning in itself, when strictly examined, but tends directly to evade or weaken the force of some important truths upon the mind—a mode of expression by which *Nature*, personified, is made to do every thing, while the Great Author of nature is never mentioned or alluded to. Surely no well informed mind can now imagine that the chain of causes and effects, which we contemplate in Natural philosophy, could ever arrange and move itself; that the material world, in which we dwell, and over which we ourselves have such dominion, was originally produced without design or impulse—or that it is without beginning, and will never have an end!

The fading leaves of the tree which I now behold from my window will, in the course of a few weeks, have fallen to the earth, and their elements will have mingled in part with the soil, in part with the atmosphere. It is in the *nature* of vegetable matter thus to decay, when separated from the unknown principle which gave it organization. In a few months, other leaves, now concealed in the buds, together with other branches, will have unfolded themselves, adding to the total bulk of the root, stem, &c which now compose the tree. It is the *nature* of trees thus to increase in bulk, and extend their p rts, by assimilating to themselves the elements contained in the earth and air. The tree with its new set of leaves will however be the same tree, though it will have changed a part of its substance : this, indeed, it has been doing ever since it first sprouted from the seed. The tree, then, was in the seed before it grew ; it is a part of the *System of nature*; and the best account we can give of its origin in common language is, that it is the *nature* (natura: that which we expect *to be brought forth*) of the seed, thus to germinate in the moist earth, and of the tree, thus set growing, to increase to perfection ; and lastly, to form in itself other seeds, capable under circumstances which will always occur in the course of *Nature* (natura rerum : that which from our knowledge of the earth and seasons, we expect will be the concurrence of events) to continue the species.

In this account of some familiar natural effects, the word *nature* has been used in its proper acceptation ; in the sense which, unless I am greatly misled by its etymology, the inventor of the term intended for it. But were I now to proceed to say that all this takes place, because *Nature* thus works, or because *she*

wills it, it would be but to run away from a plain
and positive account of the matter, already on record,
to a notion which is at best very obscure and indefi-
nite. I might indeed imagine the existence of a power
or principle, distinct from Omnipotence, and super-
seding the necessity of creation and Providence, subsisting in matter from all eternity, and manifesting
itself in an infinite variety of forms and operations—I
say I might choose to *imagine* this, but I could never
demonstrate or render it probable. I should, then,
have nothing solid to oppose to the positive authentic
history of the matter, which is this, That " in the
beginning" (of the system of nature as we now behold
it) " God created the heaven and the earth"—that
among other provisions for the use and sustenance of
the future inhabitants, He caused the earth to bring
forth " the tree yielding fruit, whose seed was in itself, after its kind." From which " beginning," by a
succession of effects, which we can investigate and
comprehend (though the created principle of vegetable life, immediately acting on matter to produce
them be hidden from us) the " kind" or species has
been continued to this day

Divine revelation was alone competent to furnish
us with just conceptions, on points of knowledge,
neither attainable by the observation of nature, nor
demonstrable by just inference from its phenomena:
and without this, it is difficult to conceive how the
idea of a spiritual energy, pervading and governing
matter, could ever have been formed by man. We
have accordingly in the book of Genesis an account
of the origin of Nature, which, while it stoops to the
simplicity of the human mind, in its ignorance of
physical science, is yet fraught with the substance of
the sublimest truths that are attainable, in the sincere

Meteorological Observations

MADE AT

TOTTENHAM, near LONDON,

IN THE YEARS

1817, 1818, 1819.

(First Published Monthly in Thomson's Annals of Philosophy).

TABLE CXXVII.

1817.		Wind.	Pressure. Max.	Min.	Temp. Max.	Min.	Hygr. at9a.m.	Rain, &c.
1st Mo. L. Q.	Jan. 10	Var.	30·58	30·40	30	20	82	
	11	E	30·40	30·11	33	25	95	
	12	SW	30·11	29·73	37	32		
	13	S	29·73	29·47	38	32		5
	14	Var.	29·47	28·90	39	27		—
	15	Var.	29·31	28·75	33	19		—
	16	SW	29·31	28·83	39	24		98
New M.	17		29·00	28·83	42	37		7
	18	S	29·03	28·99	45	37		
	19	S	28·99	28·79	45	41	58	7
	20	SW	29·42	28·79	45	31	66	—
	21	SW	29·81	29·71	44	31	77	—
	22	SW	29·80	29·79	48	44	80	9
	23	SW	30·12	29·80	52	45	73	—
	24	SW	30·25	30·12	50	45	87	—
1st Q.	25	SW	30·25	30·16	52	45	78	—
	26	SW	30·26	30·10	46	39	76	
	27	Var.	30·38	30·31	50	39	92	6
	28	E	30·25	30·22	43	40	70	—
	29	Var.	30·29	30·24	47	37	75	—
	30	N	30·37	30·20	51	40	96	
	31	NW	30·46	30 33	52	32	80	
2d Mo.	Feb. 1	NW	30·44	30·41	49	34	75	
Full M.	2	W	30·41	30·27	45	37	92	
	3	SW	30·27	29·95	41	38	70	
	4	SW	29 69	29·50	43	35	57	2
	5	NW	29·90	29·69	50	38	73	4
	6	W	30·09	29·90	54	40	62	
	7	W	30·21	30·09	51	42	63	
			30.58	28.79	54	19	76	1·38

NOTES.—First Mo. 10. Fair: hoar frost: misty. 11. Much rime: very red *Cirrostrati* at sun-rise: in the course of the day the rime mostly came off the trees, with a SW wind. 12. Grey lofty sky: *Cirrocumulus* p. m. 13. Misty: some rain after dark. 14. Clear, a. m. with *Cirrostratus* to S: from whence afterwards came on cloudiness. 15. A considerable fall of snow from SE, followed by sleet: snow at intervals, with a moderate breeze: clear frost at night 16. Misty, gloomy, a. m. the wind very light, S: then a steady breeze, SW, and decided thaw, with much sleet and rain: the product of the rain-guage is that of the guage at the laboratory, my own having been accidentally overfilled. 17 The wind, for the first time in this period, blew a moderate gale in the night. 18. Fair day: somewhat windy night. 19. Fair: the wind E, with a lofty overcast sky, and much *scud*: at noon an electric-looking compound state of the clouds: after dark, rain from the southward, and a hard gale by morning. 20. Fine day: rain after dark: windy night. 21. Very fine day: a stiff breeze, with summer-like clouds in a blue sky: *Cirrostratus* at sun-set, and a lunar corona: windy night, and a dash of rain towards morning. 22. Drizzling at intervals: a gale at night. 23. Windy: a little rain, p. m.: at night a moderate gale. 24. *Cirrocumuli*, a. m. well formed from plumose *Cirri*: afterwards a pretty sudden obscuration, and some dripping. 25. Overcast: misty: a very little rain. 26. Ten minutes' sun about noon: the blackbird and robin sing much. 27. Misty and cloudy, as heretofore: at night the wind E, with moonlight and flying clouds. 28, a. m. Small rain: gloomy. 29. *Cumulostratus*: some sun at mid-day: at night wind N, with a veil of *Cirrostratus*. 31. Misty morning, followed by a very fine day, with *Cirrus* and *Cirro-cumulus*: the hygrometer receded to 52°.

Second Mo. 1. Hoar frost: a fine day with a gradation of clouds from *Cirrus* to *Cumulostratus*, ending in an overcast sky. 2. Grey sky. 3. Misty: cloudy. 4. *Cumulus* and sunshine: at evening, thick to the SW: the wind rose to a moderate gale, with a shower. 5, a. m. High wind and clouds: dripping at night. 6. *Cirrostratus*: windy. 7. A fine sky of *Cirrocumulus*: windy, especially at night. The surface is considerably dried of late, and the roads tend to be dusty.

RESULTS.

Winds, with little exception, westerly: from the new moon to the first quarter, a SW wind, which was uniformly moderate by day, and in-creased in force in the night.

Barometer: Greatest height .... 30·58 in.
　　　　　Least .... 28·79 in.
　　　　　Mean of the period............ 29·846 in.
Thermometer: Greatest height.............. 54°
　　　　　Least....................... 19°
　　　　　Mean of the period............ 40·03°
Mean of the hygrometer.............. 76°
Rain .... 1·38 in.

The hygrometer having undergone some repair, was exposed (after adjustment) for 24 hours before the observation of the 19th, which is probably, therefore, accurate. It appears that on this day there was a tremendous gale on the coasts of Devon and Cornwall, which did much damage, particularly at Plymouth.

Paris, Jan. 23.—A storm of wind and rain was experienced on the 15th inst. at Niort, which did great damage to the surrounding country.

On the 29th of January a mist prevailed at Naples, so dense as to produce a darkness of several hours. This is a rare occurrence in that delicious clime;—(Papers).

	1817.		Wind.	Pressure.		Temp.		Hygr.	Rain,
				Max.	Min.	Max.	Min.	at 9 a.m.	&c.
2d Mo.	L. Q.	Feb. 8	NW	30·27	30·21	50°	43′	60	
		9	NW	30·29	30·10	51	39	61	
		10	Var.	30·10	29·80	50	33	77	—
		11	Var.	30·00	29·38	41	28	63	34
		12	NW	29·77	29·38	46	33	88	7
		13	SW	29·45	29·40	53	38	60	
		14	SW	29·75	29·43	53	35	83	42
		15	NW	29·66	29·43	50	35	53	3
	New M.	16	SW	29·90	29·66	47	32	79	—
		17	SW	30·02	29·90	55	45	73	—
		18	NW	30·09	30·02	54	34		3
		19	SW	30·15	29·75	46	38	63	—
		20	SW	29·50	29·38	48	33	63	6
		21	SW	29·58	29·36	44	34	60	—
		22	NW	29·90	29·58	47	33	65	
		23	W	29·79	29·68	49	39	53	4
	1st Q.	24	NW	29·95	29·79	48	38	63	
		25	W	29·69	29·62	51	40	62	11
		26	W	29·81	29·54	47	40	70	10
		27	NW	29·79	29·54	50	39	59	3
		28	SW	29·76	29·67	54	43	62	—
3d Mo.		March 1	W	29·68	29·47	53	32	58	
		2	SW	29·38	29·18	49	35	60	14
	Full M.	3	SW	29·14	28·84	50	36	59	50
		4	NW	29·24	29·14	45	30	60	—
		5	W	29·24	28·78	47	34	65	25
		6	W	29·22	28·78	43	28	62	
		7	W	29·10	28·98	46	34	63	—
		8	NW	29·40	29·10	43	28	60	55
		9	NW	29·91	29·40	45	29	61	—
				30·29	28·78	55	28	64	2.68

Notes.—Second Mo. 8. The light of an Aurora Borealis was very perceptible about ten, p. m. through the clouds which overspread the sky to the N : a windy night, with a little rain, followed. 9. Calm : grey sky, with the lighter modifications : at sun-set the clouds exhibited a splendid set of tints : close to the horizon was a clear space, lemon-coloured ; above this, crimson lights, with shadows of grey and purple, in a variety of figures, streaked, waved, and clustered ; of those in the E some were rose-red, others a tender green : a windy night ensued. 10. Fair : roads dusty. 11. Snow, a. m. with a gale

at NE: in the night a southerly gale, with rain. 12. Showers.
13. Misty: small rain: windy. 14. a. m. *Cirrostratus*: gloomy: fair
day, with clouds: windy night, with rain. 15. Windy night.
16. Windy: *Cumulus* beneath linear *Cirrus*, passing to *Cirrostratus*.
17. Cloudy: some rain morning and evening. 18. Dripping at intervals: windy. 19. a. m. calm: the dew drops frozen clear on the grass:
a very fine day ensued, with *Cumuli*, and a breeze: windy night.
20. Much wind at S this evening. 21. Fleecy *Cumuli* beneath a hazy
sky, with the lighter modifications: inosculation and *Nimbi* followed,
with rain, sleet, and snow. 22. Fair: sun and clouds. 23. Windy:
shower at night. 24. A bright haze at sun-rise and sun-set. 25. Fine
day: some *Cirrostrati* assumed an arrangement not very frequent, of
discs piled obliquely on each other. 26. *Cumulus*, capped with *Cirrostratus*: lunar corona. 27. After a gale through the night, rain before
9 a. m.: *Nimbi*, with hail, p. m.: at night large *Cirri*, very conspicuous
by moonlight, stretching SE and NW. 28. Fair, save a light shower.
   Third Mo. 1. Fair: windy. 2. A trace of solar halo about nine,
in some *Cirri*, which soon subsiding went off with the wind to SE,
grouping into forms like the crown of the *Nimbus*: *Cumulostrati* suc‑
ceeded, which, p. m. gave place again to *Cirrose* obscuration, with a
southerly gale and showers at night. 3 a. m. Overcast: p. m. steady
rain: at sun-set a hazy sky, and much vapour: a highly rarefied
Cumulostratus in the SE: a hard gale, with rain, at night. 4. Pale
sky, a. m.: after which passing *Nimbi* and a little hail: calm night.
5. Hoar frost: fair, with *Cumulus* and *Cirrus*: evening, very large‑
Nimbi: shooting stars: wind. 6. Wet morning: then fair, with various
clouds: night frosty. 7. Pretty thick ice: fair day: rain at night.
8. Windy: snow in flakes about $1\frac{1}{4}$ inch diameter: sleet and rain: at
noon large *Cumuli* in the N, passing to *Cumulostrati*, the sky above
them being blue to 15° of the cyanometer: about two, p. m. a sudden
shower of hail from a dense lofty *Nimbus*: the balls were opaque, in
the form of a cone with a rounded base about $\frac{1}{2}$ inch diam., and com‑
posed entirely of striæ meeting at the apex of the cone: we have had
similar hail repeatedly of late: frost (after rain) at night. 9. The
lighter modifications prevailed, a. m. the *Cirri* pointing to NW: after
these, lofty *Nimbi* formed in the midst of groups of *Cumulus*, letting
fall light showers: the night was clear frost.

RESULTS.

Winds Westerly.

Barometer: Greatest height .... 30·29 in.
Least ... ... 28·78 in.
Mean of the period .... 29·592 in.

Thermometer: Greatest height ... . 55°
Least.... .... 28°
Mean of the period........ . 42·06°

Mean of the Hygrometer 64°: its drier extreme several times about 40°

Rain..................... 2·68 in.

On the 2d and 3d of Third month there were violent thunder-storms to the W and S, the latter of which came as near to us as Tunbridge, but neither of them was much perceived here, save in the evident electric state of the clouds on the latter evening.

Amsterdam, Feb. 27.—It has blown hard for several days past, but in the night of the 25th it became a perfect hurricane: yesterday it abated, and is now moderate weather.

Thunder Storms.

During the night between the 26th and 27th of February, there was a hurricane at Glasgow, accompanied with rain, hail, thunder, and lightning. The wind was extremely violent, the thunder awfully loud, and the deep red flashes of lightning cast a glare during the whole night. The hail stones broke the windows in all directions.

A storm, of singular awfulness, raged over the city of Dublin, the whole of Thursday morning last, (February 27th,) accompanied with loud peals of thunder, frequent and vivid lightnings, and the heaviest showers of hail and rain. Alternate intervals of calm succeeded every blast of the tempest, which was at its height at four o'clock.

Aurora Borealis.

Edinburgh, March 6.— A little past eight o'clock, p m. a beautiful *Aurora Borealis* nearly resembling that which appeared in September last, was distinctly visible here for a considerable time. A similar beautiful arch of bright light stretched across the heavens. It sprung from a point nearly ENE, and, passing the zenith, terminated in the opposite point of the horizon. Its eastern limb was the brightest and best defined. The horizon in almost every point was obscured by dark broken clouds, which rendered both its beginning and termination less distinct than the last.

Sunderland.—On the 3th March, about seven in the evening, during a strong gale from the NW, which had continued five days, was observed here a most beautiful *Aurora Borealis.* It began in single bright streamers in the N, and NW, which gradually increasing, covered a large space of the hemisphere, and rushed about with amazing velocity and a fine tremulous motion.—About eleven o'clock, part of the streamers appeared as if projected from a centre south of the zenith, and looked like the pillars of an immense amphitheatre, presenting the most brilliant spectacle that can be conceived, and seeming to be in a lower region of the atmosphere and to descend and ascend in the air, for several minutes.—*R. Pensey in Thomson's Annals, vol. 9, p. 250.*

Inundations.

From the Mayne, March 7.—For some days past, the waters have risen in a terrible manner on both sides of the Rhine, and that river itself has reached an uncommon height. It is about the same as it was at the end of the year 1800. The Kinzig and the Schulter are, however, not quite so destructive as they were two months ago.

The alarm bell has frequently been sounded in the communes about Kehl, in order to collect people to strengthen the Rhine-dikes. Hitherto the danger has been averted; the bridge of boats at Kehl is still standing, and the communication between the two banks open: but from the left bank we have the most melancholy accounts. The rivers there have every where overflowed their banks. Great ravages have been caused about Strasburg by the Ill and Breusch. All the fields and gardens round Strasburg form one great lake. All the streets near the river are under water, and the communication kept up by boats. The Ill has done still greater damage about Schlettstadt. Several persons and a quantity of cattle have perished.—*Allgemeine Zeitung, March* 12.

Genoa, March 15.—They say here—there has been no winter south of the Alps, this year. As we passed through France, we often saw the country inundated from the snow and rains, but chiefly the latter. In Piedmont, the dust was the only thing that troubled us.

In addition to such notices as the above of the swollen state of the rivers in some parts of Swabia, France, &c. the foreign papers detail a number of accidents by *Avalanches* in the Tyrol, the Grisons, and Swisserland. The reader will perceive that the occurrence of this kind of disaster is perfectly consistent with that of inundation. The same excess of snow and rain, which loosens these overwhelming masses from their bed, must also tend to overcharge the natural channels, in its escape through the lower country. Whether the slight earthquakes which, it appears, have been frequent of late on the Continent, may not also contribute to the production of avalanches, is a question for observers on the spot to decide.

The Spring in France.

Paris, March 2.—The drivers of the little carriages for Versailles call out ' *There is still one seat left for Versailles; come and see the spring at Versailles.*' The crowd of curious persons going thither is very great. The fact, which gives occasion to this, is the fine sight presented by the trees in the great Park, which display a vegetation, such as is seen in the month of May. Several trees in the Park of Trianon are covered with new leaves; the hawthorns in the open air are loaded with flowers. May the hope, which this early spring gives us, not be destroyed by frosts in the month of May! The Chronicle of Bullinguer mentions, after a calamitous year, the summer of 1540. The fine weather and the heat lasted from the month of February to the 19th of September, and during all this period it rained but six times. At the end of May ripe cherries were eaten, and grapes in July; the 25th of June was the midst of the harvests; and at the beginning of September, the vintage was at its height. Bullinguer adds, that this year was equally remarkable for the extreme abundance of wine, corn, and all sorts of fruit.

Note.—Subsequent events have shewn that the writer of this was a year too early in anticipating a parallel to the summer of 1540: it occurred in 1818!

State of the Winds to the Southward.

About 40 sail of outward-bound vessels are lying at the Motherbank, Spithead, and Stokesbay, waiting a fair wind to proceed on their respective voyages: many of them have received their Custom-house clearances ten weeks. Though we have had such a continuance of westerly winds, yet it is stated by the master of a vessel from Teneriffe, that he experienced *nothing but northerly and easterly winds* on his voyage, until his arrival in Channel soundings. The Agricola, Captain Tabor, of and for New York, with passengers from Portsmouth, put into St. Helen's on Saturday, with the loss of her sails. She sailed 25 days since (having been previously detained there two months by contrary winds), and during that time she has never been able to get so far to the westward as Plymouth.—*P. Ledger, Mar.* 11.

TABLE CXXIX.

1817.	Wind.	Pressure. Max.	Min.	Temp. Max.	Min.	Hygr. at 9a.m.	Rain, &c.
3d Mo. L. Q. March 10	NW	30·10	29·91	47	28	75	
11	SW	30·10	29·94	53	39	66	—
12	W	29·91	29·84	55	47	70	1
13		30·15	29·91	55	42	65	
14	NE	30·22	30·15	51	30	73	
15	SE	30·22	30·16	50	27	70	
16	E	30·23	30 16	49	27	82	
New M. 17	SE	30 24	30·20	48	25	85	
18	Var.	30·20	29·85	52	33	50	
19	NW	29·75	29·74	47	27	52	
20	N	29·88	29·75	34	24	47	
21	N	29·90	29·86	39	17	59	
22	SE	29 97	29·90	39	19	80	
23	SW	29·92	29 88	46	24	58	—
24	W	29·88	29 72	55	39	80	—
25	Var.	29·85	29·72	58	34	74	7
1st Q. 26	Var.	30·00	29·85	52	34	52	13
27	Var.	30·05	29·92	44	27	62	
28	SW	29·92	29 76	50	38	72	4
29	W	29·99	29·76	55	45	60	—
30	NW	30·23	29·99	59	39	50	
31	NW	30·51	30 23	54	32	64	
4th Mo. Full M. April 1	S	30·51	30·37	56	36	67	
2	SE	30·37	30·27	58	33	50	
3	E	30·33	30·27	60	37	65	
4	NE	30·33	30 30	56	34	70	
5	N	30·32	30·25	53	26	64	
6	NE	30·43	30·32	50	37	52	
7	E	30·37	30·20	52	30	60	
		30·51	29·72	60	17	64	0.25

NOTES.—Third Mo. 10. Fine, with *Cumulostratus.* 11. A mist, probably from the Thames, there having been much *Cirrostratus* at sun-rise in the SE: cloudy, p. m. with a few drops. 12, a. m. *Cirrostratus* in flocks: at evening a slight shower, with wind. 13. Fair: overcast with *Cumulostratus.* 14. This morning at eight the wind sprang up at NE, a gentle breeze, which, being propagated upwards, carried a veil of *Cirrostratus* off to SW: in the evening the sun's disk was curiously disfigured by the intervention of *Cirrostrati,* with vapour: after being divided, and afterwards crossed as by belts of this cloud, the lower portion came out much extended horizontally, while the part yet obscured became somewhat conical upwards. 15. a. m. *Cirro-*

stratus: misty to SW, after which light breezes and general cloudi‑ ness. 16. Hoar frost: fair: wind SE a. m., NE p. m. 17. A drip‑ ping mist, after hoar frost: then *Cumulus*, and the wind S. 18. Hoar frost, misty morning, SE: clear day: p. m. the wind SW, a smart breeze: clouds after dark. 19. a. m. Wind SW: *Cumulus*, beneath *Cirrocumulus* and *Cirrostratus:* p. m. windy at NW: *Cumulostrati* and *Nimbi*, with a little hail. 20. A gale at NNW, tending continu‑ ally to go to N: a very scanty snow at intervals. 21. Very fine: *Cumuli* prevailed, which evaporated at sun-set: the roads quite dusty: wind tending to E, a smart breeze: night calm. 22. Hoar frost: hygr. noted at eight a. m.: very light breeze. 23. Hoar frost: fine day: evening obscured by *Cirrostratus*, which descended from above. 24. Some drizzling rain this morning. 25. Hoar frost: rain: a hail shower: p. m. the wind NE. 26. a. m. Overcast with *Cirrostratus:* small rain, p. m. 27. Very fine day: wind a. m. NNE, with *Cumulus* and *Cirrostratus*. 28. Wind S, a. m. with *Cirrostratus:* drizzling rain. 29. Temp. 50° at nine, a. m.: windy at SW. 30. Very fine morning, with dew: *Cumulus* beneath *Cirrocumulus* and *Cirrostratus:* a few drops of rain: a small yellow lunar halo: much wind in the night. 31. Windy: *Cumulus* beneath large *Cirri:* a lunar halo, white and of large diameter.

Fourth Mo. 1, 2. Light driving mists, followed by fine days. 3. Hoar frost: rose-coloured *Cirri* at sun-set. 5. Cloudy: a few drops: misty night. 6. Hoar frost: *Cumulus*, with *Cirrocumulus:* windy. 7. Windy at SW by night, with mist.

RESULTS.

Winds for the most part light and variable, but on the whole Northerly.

Barometer: Greatest height .. .. 30·51 in.
Least .... 29.72 in.
Mean of the period..... ...... 30·07 in.
Thermometer: Greatest height .... 60°
Least .... 17°
Mean of the period.. . ... 41·5°
Mean of the hygrometer........ ...... 64°
Rain.... ... . 0·25 in.

The change from the turbid Atlantic air, which had for many months been flowing over us, to a dry transparent medium, was, from the com‑ mencement of this period, strikingly obvious to the sense. The sun assumed a splendour, and the moon a brilliancy, to which the eye had been long unaccustomed, and distant objects seemed as it were restored to the landscape. The mean of the barometer is the highest that has occurred to me since the spring of 1813: the *ten* dry days about the commencement of the period were the first that had happened in strict succession for twelve months; and there has not fallen so little rain in any lunar period that I have registered since the beginning of 1810. The evaporation has doubtless been excessive, and I regret that I have kept no account of it: for the state of the hygrometer did not fully indicate the dryness of the air, on account of the misty mornings.

B

1817.		Wind.	Pressure. Max.	Min.	Temp. Max.	Min.	Hygr. at 9 a.m.	Rain, &c.
4th Mo. L Q.	April 8	Var.	30·20	29·89	58°	34°	59	1
	9	NE	29·97	29·89	46	28	46	—
	10	N	30·23	29·97	40	25	50	—
	11	NW	30·23	30·02	46	29	53	
	12	NW	30·01	29·96	49	39	63	
	13	NW	30·04	30·00	55	38	53	6
	14	NW	30·00	29·93	60	42	55	—
	15	NW	29·72	29·67	61	41	50	—
New M.	16	N	30·11	29·72	48	32	52	2
	17	N	30·30	30·11	42	34	43	
	18	N	30·37	30·30	53	26	50	
	19	NE	30·32	30·30	55	40	42	
	20	NE	30·34	30·33	55	34	50	
	21	NE	30·34	30·27	59	32	46	
	22	SE	30·27	30·17	57	29	60	
	23	NE	30·20	30·17	50	27	59	3
1st Q.	24	NE	30 20	30·14	52	35	52	2
	25	NE	30·14	30·12	44	36	46	
	26	NE.	29·93	29·87	49	40	44	
	27	Var.	30·09	29·93	50	32	40	
	28	W	30·09	29·91	58	43	45	
	29	NW	29·91	29·70	48	37	47	—
	30	NE	29·81	29·69	50	39	55	·10
5th Mo. Full M.	May 1	NE	29·93	29·81	48	34	50	
	2	NW	29·93	29·84	56	30	42	
	3	W	29·77	29·72	60	45	45	—
	4	NW	30·01	29·77	60	32	41	
	5	SW	30·06	29·95	64	35	48	
	6	N	30·16	30·06	64	36	34	
	7	SE	30·06	29·77	60	37	50	4
			30·37	29·67	64	25	49	0·28

NOTES.—Fourth Mo. 8. The wind was for some time at SW: rain in the night. 9. Cloudy, a.m.: a shower of driven granular snow in the night. 10. *Cumulostrati* and *Nimbi*, giving small quantities of snow. 11. *Cumulostratus*: windy. 12. Mostly overcast: very light rain at intervals. 13. Small rain, a.m.: fair, p.m. 14. A little light rain. 15. Fair: large plumose *Cirri* above *Cumuli*. 16. a.m. A strong gale from NW and N, with a shower and hail: rainbow: fair day after. 17. *Cumulostratus*: dark sky: windy. 18. *Cumulostratus*: the wind veers to NE and NW: calmer day. 19. The hygrometer noted at ten:

Cumulostrati prevailed, surmounted by the lighter modifications : windy : the part of the moon's disc in shade distinctly visible, and the light crescent very conspicuous in the evening: a small meteor passed to the NE. 20. a. m. Windy, not steady to NE : *Cumulostrati.* 21. a. m. *Cirri* pointing westward, with *Cumuli* beneath : afterwards an arrangement of this cloud in regular parallel streamers from NW to SE,, which became red at sun-set. 22. With the SE wind this morning the *swallows* appeared, but few in number, and flying feebly : a serene evening, after *Cumulus* and *Cumulostratus.* 23. Hoar frost early : cloudy : windy : a shower from NE, p.m. : clear evening : the hygrometer to-day receded to 32°, and the superior part of the clouds, after the rain, presented a configuration like the *pores of sponge*, which I have not observed before for some years. 24. *Cumulostratus* : windy : a shower at night. 25. a. m. Overcast : windy : *Cumulostratus.* 26. The same, the breeze growing stronger. 27, 28. Chiefly overcast with *Cumulostratus* and large *Cirrocumulus.* 29. The same, with *Cirrostratus* : a slight shower by night. 30. A moderate gale at NE, with showers and much cloud : *Nimbi* : a little hail.

Fifth Mo. 1. Cloudy : windy. 3. A slight shower in the night. 5. The hygrometer receded to 32°. 6. The wind went from N to E. 7. Wind SE : a breeze : very clear all day, and a full orange twilight: by six, a. m. the 8th, it was however SW, with a slight shower.

RESULTS.

Winds almost uniformly northerly, and moderate in force.

Barometer : Greatest height ... . 30·37 in.
Least .... 29·67 in.
Mean of the period .... 30·028 in.

Thermometer : Greatest height .... 64°
Least .... 25°
Mean of the period .... 43·85°

Mean of the hygrometer .... 49°
Rain... .... 0·28 in.

Vegetation has been peculiarly slow during this period.

On the 19th of Third Month last I had an opportunity of observing that rare phenomenon the *Anthelion.* It was formed on the perpendicular part of a lofty dense *Cumulostratus*, which happened to present in the NE at near five, p.m. a surface directly opposed to the sun, reflecting *an image of the disk*, at the same apparent height from the horizon. In a few minutes, and almost as soon as I had satisfied myself of the fact, it was obliterated by a new protuberance in the cloud destroying the direct reflection. An Anthelion observed by Swinton near Oxford in 1762 is described, with a figure, in Vol. XI. of the Phil. Trans. Abridged, p. 532, to which the reader is referred ; but in the present instance, the whole cloud being bright, the contrast between the general surface and the sun's image was probably less striking than in Swinton's observation.

TABLE CXXXI.

	1817.	Wind.	Pressure. Max.	Min.	Temp. Max.	Min.	Hygr. at9a.m.	Rain, &c.
5th Mo.	L. Q. May 8	Var.	29·72	29·68	72°	42°	49	8
	9	NE	29·68	29·49	54	37	59	
	10	SW	29·43	29·35	60	38	45	—
	11	Var.	29·35	29·23		43	48	15
	12	W	29·63	29·23			45	—
	13	SW	29·63	29·59	59	34	58	13
	14	W	29·74	29·59			41	45
	15	SW	29·90	29·74	62	33	49	
	New M. 16	Var.	29·90	29·80	65	33	50	
	17	SW	29 80	29·55	67	39	42	1
	18	SE	29·43	29 36	72	44	53	11
	19	NE	29·46	29·43	53	38	67	—
	20	NE	29·42	29·34	52	43	80	—
	21	NE	29·42	29·40	48	38	59	—
	22	SW	29·42	29·40	57	40	50	1·44
	23	W	—	—	—	—		
	1st Q. 24	NE	29·40	29·27	62	35	65	18
	25	SE	29·19	29·16	57	45	58	29
	26	E	29·37	29·17	63	41	44	
	27	NE	29·59	29·35	69	38	53	
	28	NE	29·59	29·56	59	47	77	
	29		29·80				51	—
	Full M. 30		29·90					23
	31		29.75	29·68	59	33	50	
6th Mo.	June 1	SW			63	42	41	
	2	W	29·64	29·58	64	46		
	3	W	29·59	29·45	64	52		
	4	W	29·99	29·45	65			
	5	SW	29·91	29·89	65	47	54	11
			29·99	29·16	72	33	54	3·18

NOTES.—Fifth Mo. 8. *Cirrocumulus*, mixed with *Nimbi*, a. m. after which, the cloudiness becoming general, a thunder-storm ensued, soon after four, p. m.: it came from the SW, with the wind at SE. 9. Cold wind, a. m. with a general cloudiness. 10. Overcast, a. m. with *Cumulostratus*: a few drops of rain. 11. *Cumulus, Cumulostratus,* and *Nimbus*: the wind NW and SW : rain with wind at night from the southward. 12. a. m. A westerly gale. 13. Showery, with hail twice. 14. Showery : hail, pretty large, at noon from the southward. 15, 16. Fair. 17. A shower, p. m.—Travelling in the interval from the 13th to the 17th inclusive as far as Leeds, in Yorkshire, and home again, I found cloudiness from large *Cumuli*, &c. general, but met with very little rain. On the 15th, passing between Leeds and Pontefract, there was a fine display of *Nimbi*, one of which let fall a heavy shower on the latter place and its environs. On the 17th, after a deep orange tint in the morning twilight, the sun rose *red* behind a *Cirrostratus*; in emerging from which the brilliant part of the disc was divided by a well defined line from the lower and coloured portion. 18. Cloudy, a.m.: gentle rain, p.m. 19. Windy, cloudy, a.m. wet, p.m. 20, 21. Rainy. 22. Cloudy. 23. 24. Some showers: a *Stratus* at nine, p.m. the latter day. 25. Thunder at a distance: showers, a.m. 27. A thick fog at night, undoubtedly from a *Stratus*.

RESULTS.

Winds variable, but for the most part westerly.

Barometer:	Greatest height	29·99 in.
	Least	29·16 in.
	Mean of the period	29·533 in.
Thermometer:	Greatest height	72°
	Least	33°
	Mean of the period	50·70°
Mean of the hygrometer		54°
Rain		3·18 in.

I had anticipated a *third* dry period, similar to the two we had experienced, and expected that the rains would return after the summer solstice: in this I have been happily mistaken. In the beginning of the present period the weather took a new type with us, the westerly current coming in again, with some discharges of electricity, bringing rain, which gradually became more plentiful, and proved exceedingly seasonable. Vegetation has passed, in consequence, from a starved and backward state, to one of considerable luxuriance and promise. It is observable that the barometer during this period has scarcely passed the boundary of 30 inches in elevation, and has certainly not descended below 29 inches. The mean temperature, though 6° higher than that of the period immediately preceding, is low for the season.

GREAT RAIN AT STUTGARD.

Paris, June 5.—Letters from Stutgard, dated May 28th inform us, that on the 26th the rain began to fall in torrents, and did not cease for three days and three nights; all the rivers were overflowed, and it produced an inundation far surpassing any thing ever recollected in that country.

The lower part of the small town of Constatt, a league from Stutgard, was under water, and the suburbs were evacuated by the inhabitants. Several individuals and many animals were drowned at Constatt: on the evening of the 28th the rain ceased.—*Pub. Ledger.*

Note.—It is observable that this unusual fall of rain in a part of Germany 450 miles SE of the Thames, began at the precise time when after nine days of wet weather it became fair with us; and that during its continuance, with a sufficiently low state of the Barometer, *we* were without rain.

1817.			Wind.	Pressure. Max.	Min.	Temp. Max.	Min.	Hygr. at 9a.m.	Rain, &c.
6th Mo.	L. Q.	June 6	SW	29·89	29·84	74	55	45	
		7	SW	29·84	29·64	77	51	50	
		8	SW	29·79	29·64	67	44	61	—
		9	SW	29·79	29·67	60	54	54	70
		10	NW	29·94	29·67	67	40	47	6
		11	SW	29·89	29·68	71	50	54	
		12	SE	29·68	29·52	—	—	54	
		13	Var.	29·37	29·17	65	48	48	39
	New M.	14	NW	30·05	29·37	61	43	41	1
		15	W	30·23	30·05	63	34	43	
		16	E	30·23	30·00	70	40	42	
		17	SE	30·00	29·62	76	50	42	
		18	SE	29·67	29·62	79	52	42	
		19	SE	29·75	29·67	83	53	48	
		20	SE	29·83	29·75	83	59	47	
		21	NE	30·00	29·83	86	59	51	
	1st Q.	22	NE	30·00	29·90	84	56	50	
		23	N	29·90	29·87	84	59		
		24	W	29·90	29·87	82	58	47	—
		25	W	29·92	29·80	77	52	61	
		26	W	29·80	29 55	76	58	42	—
		27	NE	29·65	29·50	83	56	43	—
	Full M.	28	W	29·92	29·65	72	45	42	—
		29	SW	29·92	29·65	74	55	40	1·07
		30	SW	29·77	29·65	68	44	39	—
7th Mo.		July 1	SE	29·65	29·37	60	55	51	
		2	SW	29·90	29·60	68	48	50	
		3	W	29·90	29·55	70	54	54	
		4	SW	29·55	29·48	68	51	44	
		5	NW	29·67	29·50	70	49		58
				30·23	29·17	86	34	47·5	2·81

NOTES.—Sixth Mo. 7. Much *Cirrocumulus*, a. m. in beds at a considerable elevation : in the evening a group of thunder clouds in the S and SE, which passed, after a single peal of thunder, to the eastward. 8. Windy : light showers, a. m. : heavier rain, p. m. 9. Stormy wet day and night. 10. Showers : *Cumulostratus* at sun-set. 11. Fine morning. 13. Wet, blowing day : stormy night. 14. Much wind and cloud, a. m. : slight shower : evening more settled. 15. Windy at NW, a. m.: *Cumulus*, with *Cirrostratus* : *Cumulostratus* : fair : *Stratus* at night. 16. Fine : *Cirrus* at evening. 17. A *Stratus* visible at four, a. m.: very fine day: luminous twilight, with the moon conspicuous: *Cirri* after sun-set. 19. Hot sun-shine: fair. 20. Lightning this evening. 21. *Stratus* at night. 22. Continued thunder in the SE, p. m.. 23. Rather cloudy, a. m.: a fine breeze: 24. Morning cloudy, then fine : in the evening, heavy rain, with hail, thunder, and lightning: hygr. before the storm at 36°. 25. Cloudy morning. 26. Misty morning: drizzling rain, then fine. 27. A thunder storm between six and seven in the evening: very heavy rain, with thunder and lightning. A waterspout passed within view to the N, of which see the account annexed. 28. Heavy showers, evening. 29, 30. Cloudy, with showers.

Seventh Mo. 5. Thunder in a mass of clouds to the south and south-west: some rain with us.

RESULTS.
Winds variable.

Barometer:	Greatest height	30·23 in.
	Least.................... ..	29·17 in.
	Mean of the period............	29·751 in.
Thermometer:	Greatest height	86°
	Least......................	34°
	Mean of the period....	61·83
Mean of the hygrometer........		47·5°
Rain...		2·81 in.

A Waterspout.

On the 27th of the sixth month, about seven in the evening, there occurred in our neighbourhood an undoubted exhibition of that rare spectacle (to observers on land)—the *waterspout.* I shall give the observations of two of our workmen at the Laboratory who saw it from Stratford, passing on their N horizon from NW to NE. I was absent myself in the West of England ; but my friend John Gibson witnessed the latter part of the phenomenon.

The weather over head had been exceedingly dark and threatening, and there had been thunder and rain in that direction ; but at the time of the observation a clear sky was discernible beneath the clouds. From a dense cloud, the base of which might be at an elevation of 20°, there issued suddenly a *descending cone,* which one of the observers compared to a steeple inverted: this returned back to the cloud : a second and a third followed, one of which came lower, with a considerable *perpendicular oscillation,* and at length *opened out* below :

and a *straight column*, which he compared to a dart, proceeded from its enlarged extremity to the earth, being visible also as a *denser body* pretty far up into the cloud. In a little time this cone also, losing its appendage, was drawn up again, and another or two, similar to the first mentioned, succeeding, closed the train of appearances, the whole having lasted about 15 minutes.

The course of this spout appears to have been over the country about Hampstead. In a communication, by another observer, to the Philanthropic Gazette, a person is stated to have been overtaken by it on Hampstead Heath, and to have been drenched by a fall of rain in very unusual torrents during its short passage. He conceived the spout to touch the top of the tree under which he had retired for shelter. The denser column seen by the observer at Stratford to proceed from the cloud admits of an explanation when connected with this fact. It was probably an extremely heavy shower, or rather *stream of water* (of small diameter compared with showers as they usually fall) generated in the axis of the cone of cloud by the strong electrical action which produced the latter, and serving ultimately as a conductor, through which the electricity rushed at once, and the equilibrium was so far restored as that a second discharge in this way could not be effected. Had it been at sea, the tendency of the superinduced moveable surface of the waters to unite with the cloud, would probably have raised up a column of salt water to meet it : and the appearances would then have made the phenomenon complete in all its parts.

The character of this period was certainly highly electrical : a display of excessive heat for ten days about the solstice, was introduced by SE winds, and ended, as usual in these cases, in a copious rain.

THUNDER AND HAILSTORMS.

The Papers make mention of violent thunder storms, attended with large hail, on the 9th of the Sixth month at Dunkeld (Scotland), and on the 10th, in the forenoon at Edinburgh and Dundee, and at midnight at Cupar in Fife. On the 24th there was a thunder storm at *Dublin*, which is a rare occurrence there.

The Bath Paper says—" The heat of the sun, during the last few days, has been more excessive than we have experienced for some years past. Friday (June 20th) Fahrenheit's thermometer stood at 82°; on Saturday, in the shade, it stood at 86° from three to five o'clock in the afternoon, being 10 degrees above summer heat, and, notwithstanding the cloudiness of the day, the thermometer was at 103 in a more exposed situation; and on Monday in the sun it rose as high as 113. At Weymouth on Saturday, the thermometer stood at noon in the shade at 86, and in the sun at 112: while at Gloucester, on the same day, it rose to 103 in the shade. This intense heat, as might be expected, has been productive of thunder-storms in many parts of the country. The vicinities of Gloucester and Tewkesbury experienced some of their effects on Saturday afternoon; and the inhabitants of the latter place have sustained considerable loss by the hail, which broke many windows.—At Lyneham, near Chippenham, on the same day, a water-spout inundated a considerable quantity of land, and occasioned a rapid rise of the Avon, which very sensibly affected the river at this distance.—Salisbury and its neighbourhood appears to have received the brunt of the storm. It commenced there about two o'clock in the afternoon, with almost an instantaneous darkness, and a violent rushing of wind from the north-east, accompanied by sheets of water and large pieces of ice. About three the wind from the north-east ceased, and suddenly commenced blowing from the south-west, with such torrents of rain for more than half an hour, that every street was flooded, and the water ran through many of the houses. The lightning was not very vivid, nor was the thunder extremely loud; but they continued during the whole of the storm, which lasted till six o'clock. Forty sheep, of a flock belonging to Mr. Swayne, of Langford, were struck down by the lightning, and six sheep and six lambs killed; several large trees were also blown down during the storm at Durnford and West Harnham. The storm also visited this city about two o'clock, and continued till five, but its effects were not marked by any extraordinary circumstance. But very considerable damage was sustained by the unexampled violence of a storm which occurred here on Monday afternoon (June 23d): the rain fell in such torrents that the common sewers were soon choaked, and the lower apartments of many houses were in consequence flooded; almost every hot and green-house, and sky-light, in the neighbourhood, suffered in its glass, more or less, from hailstones, many of which were two inches in circumference."—*Pub. Ledger, June* 27.

I had occasion from a curious accident to notice the weight of rain on the day of the last mentioned thunderstorm. Travelling with a relation, we had entered our inn at *Exeter* just as the storm began, it being evening, and taken possession of a parlour on the first floor; when in an instant a copious stream of sooty water flowing from the fireplace (the gutter having by some means found a discharge into the chimney) compelled us to summon the servants, and retreat hastily into another apartment.

The Dutch Papers contain various accounts of severe damage done by the storms of hail, accompanied with thunder and lightning, on the 25th, 26th & 27th of June.

Boitzenburg, July 2.—Yesterday we had a tempest with a terrible hailstorm, which has almost entirely laid waste the fields of Horst and Viechof, in Lauenburg; so many windows are broken, that there is not glass enough in town to repair them; the hail was jagged, and many pieces above one inch long.—*Pub. Ledger.*

From the Quebec Papers of the 10th of July.

FURTHER PARTICULARS OF THE EARTHQUAKE AT ST. JOHN'S, NEW BRUNSWICK.

The Earthquake was felt on the 22d ult. over all the island of Grand Manan, and has been thus described to us:—The reporter was awakened just after daybreak by the shock of a loud sound, and a violent shaking of the house at the same instant. The shaking ceased very soon; but the sound, he thinks, continued from 30 to 45 seconds after he awoke, gradually lessening till it entirely died away. Some of the inhabitants say it was perceived much longer, but the best opinions were, that it lasted a full minute. All agree in describing the motion as most violent, and the sound to have been very loud; the weather at the moment was fine and serene, with a light breeze of wind from the northward: the previous day it had been uncommonly hot for the season. During the 22d, the weather continued fine and warm, the wind easterly, and light. This earthquake we already trace from Boston to Portland, St. Andrew's, and Frederickston, nearly 400 miles: and in another line, of a similar distance and parallel direction, taking the opposite side of the Bay of Fundy in its route, and going through Grand Passage, Digby, Annapolis, and Windsor, extending in each end of this line, from which we have yet no tidings.—*Pub. Ledger.*

1817.		Wind.	Pressure.		Temp.		Hygr. at 9 a.m.	Rain, &c.
			Max.	Min.	Max.	Min.		
7th Mo. L. Q.	July 6	SW	29·68	29·59	70°	45°	48	—
	7	SW	29·73	29·67	71	46	45	
	8	SW	29·78	29·73				
	9	SW	29·78	29·73	75	47	44	
	10	S	29·68	29·66	76	48	46	—
	11	S	29·73	29·66	73	55	55	—
	12	NW	29·88	29·73	68	48	43	
	13	SW	29·88	29·60	71	50	49	13
N. M.	14	NW	29·60	29·14	70	52	65	72
	15	SW	29·54	29·06	70	49	46	18
	16	NW	29·77	29·54	69	49	46	
	17	NW	29·77	29·73	71	52	48	15
	18	NW	29·85	29·77	67	46	44	
	19	NW	29·90	29·85	69	42	41	
	20	S	29 90	29·81	72	54	61	3
1st Q.	21	SW	29·85	29 70	73	56	63	4
	22	SW	29·85	29·75	70	52	40	
	23	SW	30·00	29·85	70	50	52	7
	24	SW	30·00	29·95	71	45	46	
	25	SE	29·95	29·80	72	50	40	
	26	S	29·80	29·60	66	47	50	20
	27	W	29·85	29·60	66	47	40	13
Full M.	28	W	30·00	29·75	68	48	39	10
	29	W	29·75	29·69	70	50	45	14
	30	W	29·75	29·67	71	48	45	3
	31	SW	29·75	29·67	71	44	45	
8th Mo.	Aug. 1	SW	29·92	29·75	69	41	49	
	2	SW	29·92	29·65		54	50	
	3	W	29·75	29·65	68	46	54	
	4	SW	30·00	29·67	68	50	53	3
			30·00	29·06	76	41	48	1·95

NOTE.— Seventh Mo. 6. Some rain, a. m. : windy : twilight orange coloured. 7. *Cumulostratus* : fair. 8, 9. Fair : cloudy : red sun-set. 10. Cloudy : calm : a light shower. 11. Cloudy : a light shower, a. m. 12. Cloudy : a light shower early : fair day. 13. Large *Cirri* : fine, a. m.: *Cirrocumulus* : *Cirrostratus* : windy : cloudy : shower, evening. 14. Cloudy morning, followed by several light showers. 15. Rain in the night, and a wet morning : much *Cirrostratus*, with a pretty calm air ; afterwards the *Nimbus* prevailed, with sudden showers ; and it was stormy at night. 16. Cloudy : calmer : fair. 17. Cloudy : *Cirro-*

cumulus: Cirrostratus: fair day: rain in the night after. 18, 19. Cloudy: windy: fair: a ruby-coloured twilight, the clouds rapidly dispersing at the time. 20. *Cirrostratus,* alternating with *Cirrocumulus:* then *Cumulostratus* and rain in the evening. 21. Cloudy, windy, a.m.: a fine display of *Cirrostratus* in elevated beds, passing to *Cirrocumulus.*

Eight Mo. 1. Chiefly showery for the last ten days, with thunder three times.

RESULTS.

The wind uniformly westerly, a single observation excepted, which was of short continuance.

Barometer: Greatest height .... 30·00 in.
Least .... 29·06 in.
Mean of the period ....,... 29·743 in.
Thermometer: Greatest height............... 76°
Least .... 41°
Mean of the period .... 59·32°
Mean of the hygrometer ... . 48°
Rain .... 1·95 in.

The period was throughout changeable, cloudy, and windy, the barometer fluctuating (save in one depression) between the limits of 29·5 and 30 inches. The rain fell chiefly in two distinct spaces of five days each, determined as it appears by the occurrence of New and Full Moon.

The following observation was communicated to me by my friend Thomas Forster, at Tunbridge Wells:

July 30, 1817.—11^h 30^m, p. m. A fine coloured *paraselene,* about 23° ENE of the moon. It lasted about three minutes; and then there broke out from it a tapering or conical band in the direction *from* the moon, i.e. ENE; and in a minute more the whole disappeared. Features of *Cirrostratus* were discernible at the edge of the thin cloud in which it was seen.

About six o'clock last evening, the Metropolis was visited by a dreadful hail-storm from the westward, which must have damaged innumerable panes of glass in houses having a western aspect. The storm lasted about seven minutes, and most of the hailstones which fell were as large as hazel nuts.—*Pub. Ledger, July* 29.

TORNADOES.

Extract of a Letter from Derby.—" Friday afternoon (July 11), about two o'clock, this neigbourhood was visited by the awful, but happily very rare phænomenon of a tornado. It advanced from the south-west, and first came in contact with the earth near the Depot, about three quarters of a mile from Derby, where it was most violent. It there tore up a fine ash tree by the roots, several large branches of which were carried to a considerable distance; and in its progress it took up a quantity of new hay from the grounds of E. S. Sitwell, Esq. (not less than half a ton) which was carried to an immense height, dispersed to a wide extent, and carried along with the clouds.—The storm happily passed over without doing any injury to the town. It was accompanied by very heavy rain, which brought down portions of the hay, some of which was very closely matted together, but the greater part was borne along with the clouds till quite out of sight. The vanes of All Saint's Church were observed to turn round at intervals during the storm, and although the air must have been so greatly agitated in the upper regions, it was perfectly calm below. The storm took a northerly direction, and as soon as it had passed away the sky became clear, and the air remarkably hot; soon afterwards the clouds collected again with rain for the remainder of the day. The quantity collected in Mr. Swanwick's rain gauge was one inch and a half."— *Pub. Ledger, July 21.*

Altona, July 16.—On the 11th of this month 16 or 17 houses were entirely overturned, and several others considerably damaged by a water-spout (tornado) in the village of Wattenbeck, near New Munser.

Pforzheim, July 12.—Yesterday afternoon a dreadful hail-storm fell here, and also at Ispringen and Eukingen; the hail-stones were all triangular, and as large as pigeons' eggs. Many persons were severely wounded in the hands and head; a great many windows were broken, and what is more distressing, the fine cornfields were almost entirely laid waste in a quarter of an hour.

INUNDATIONS ABROAD.

All the accounts from the Eastern part of Switzerland announce the terror and the damage caused by the late inundations.

The storms have carried desolation into the lower parts of the Canton of Glaris. The Linth has broken its dikes in three places. The bridges of Glaris and Helstal have fallen down; that of Miolis threatens to go to ruin. Gessau, Rutti, Fleriscue, and Hagelschauer in Teggenburg, felt the whole violence of the storms of the 4th and 5th (July ?). All the torrents have overflowed. The bridge of Aberglatt is in ruins.

At Basle the Rhine rose so much on the 6th, as to inundate the city as far as the fish markets; the citizens were forced to cross the streets in boats. The Rhine continually brought down with it trees, parts of buildings, and drowned animals, shewing by these numerous wrecks the ravages it has exercised elsewhere. At Constance the lake was much higher on the 6th than in 1666, and even some inches higher than in 1560. — In spite of unremitting exertions, the bridge of Lindau is carried away. On the banks of the lake many Communes are under water, and it is feared will continue so for a long time, the vents by which the waters must run off when the lake falls, being too small. In the Lower Rhinthal the surface of the waters which covers the fields and the roads, and upon which one may easily navigate between half ruined houses, was three leagues in circumference. At Horn, and all along the lake, a great many buildings are abandoned; the waters threaten the foundations of the most solid edifices. In the Oberland, many bridges have been carried away. The fields, the meadows, the plantations, were entirely submerged, and pieces of the soil were seen floating about, torn up by the fury of the waters, covered with potatoes, vegetables, and hay. On the 9th, during a violent tempest, the lightning struck the village of Deterswell, and burnt a house. Near Neutingen many cattle were killed by lightning. Other accounts, equally distressing, have been received from other quarters.—*Pub. Ledger, Aug.* 2.

WEATHER IN ICELAND.

Rikavick, Aug. 17, 1817.—Last winter was one of the severest we have had for a long while, in particular from the beginning of February to the end of March, with changeable winds and heavy snow. From the beginning of April until the 1st of May, we had often fine and mild weather with thaw; but on May 2, we had a storm from the north with much snow; and from that day until July 7 we had nothing but northerly winds with frost and cold weather. The Greenland drifting ice, which had left the northern land in the beginning of April, returned again in the first days of May, and surrounded the whole of the western, northern, and eastern land. From about July 7 the weather has been very dry and often pretty warm.—*Thomson's Annals of Philosophy*, vol. ii. p. 229 ; from the Danish Official Gazette, communicated by Sir Joseph Banks.

1817.		Wind.	Pressure. Max.	Min.	Temp. Max.	Min.	Hygr. at 9a.m.	Rain, &c.
8th Mo.	L. Q. Aug. 5	NE	30·05	30·00	70	48	47	
	6	NE	30·00	29·85	74	40	44	
	7	NW	29·83	29·45	75	55	44	
	8	SW	29·58	29·45	67	48	55	
	9	W	29·72	29 59	71	45	44	
	10	SW	29·72	29·71	71	34	50	10
	11	SW	29·72	29·33	71	48	52	—
New M.	12	SW	29·33	29·17	68	52	50	45
	13	SW	29·55	29·33	67	54	55	
	14	W	29·61	29·55	71	54	62	3
	15	W	29·78	29·55	68	46	48	4
	16	S	29·58	29·55	70	48	45	22
	17	W	29·87	29·59	66	45	50	6
	18	SW	29·64	29·59	66	54	49	28
1st Qr.	19	SW	29·64	29·55	68	54	60	6
	20	W	29 64	29·55	69	50	50	—
	21	NW	30·02	29·64	59	42	64	12
	22	E	30·02	29·94	63	35	50	
	23	SE	29·94	29·68	67	48	58	
	24	S	29 68	29·20	65	50	45	—
	25	SW	29·20	29·00	62	48	63	29
Full M.	26	S	29·08	28·90	64	44	55	15
	27	SW	29 55	29·08	68	51	48	3
	28	SW	29·64	29·54	68	51	50	12
	29	W	29·80	29·64	69	47	53	
	30	SW	29·80	29·75	71	54	53	18
	31	SW	29·95	29·75	67	41	53	
9th Mo.	Sept. 1	NE	29·98	29·95	69	37	64	
	2	E	29·98	29·83	69	48	58	
			30·05	28·90	75	34	52·3	2·13

NOTES.—Eighth Mo. 21. A wet morning: windy at N: p. m. cloudy, with wind about NNW: pretty calm at night. 22. Fair, with *Cirrostratus* beneath *Cirrus*: gold-coloured moon: calm at night. 23. Fine morning: there is said to have been hoar frost: a few *Cumuli* appeared, which soon became heavy *Cumulostratus*: and in the evening it was quite overcast, with a few drops of rain. 24. Fine, a. m.: the wind SE: a little rain, p. m.: during the day a singular anomalous veil of cloud overspread the sky, in which the *Cirrostratus* on the whole predominated: the lower surface of these clouds put on fine

crimson and grey tints at sun-set, and the lights formed by the moon
shining through them were peculiarly soft and pleasing. 25. Cloudy,
a. m.: small rain: the wind gentle, veering to S: it rained much of
the day at intervals: afterwards appeared groups consisting of *Cumulostratus* and *Cirrocumulus,* with *Nimbi:* hazy moonlight. 26. *Cirrostratus* in the morning: then *Nimbus,* and some rain: the wind gone
back to SE, and moderate: many sudden showers of small amount
from ill-defined clouds amidst haze: a bow soon after three, p. m.:
windy night. 27. Fine morning: much dew: calm: *Cumulostratus*
tending to *Cirrocumulus* above: some rain at mid-day. 28. Fine
morning: *Cumulus* passed to *Cumulostratus:* a very few drops fell,
p. m.: and there followed wind, succeeded by calm, with *Cirrostratus*
and haze: rain in the night. 29. Fair: brisk wind, with various
clouds. 30. A veil of *Cirrostratus* in flocks, a. m., with this, *Cumulus*
rapidly inosculating formed *Cumulostratus,* which was heavy through
the day: in the evening much *Cirrostratus,* succeeded by small rain:
in the night a heavy shower. 31. Fine, with *Cumuli,* carried by a
strong breeze.

Ninth Mo. 1. Misty morning, with *Cirrostratus* above, to which
succeeded *Cumulostratus.* 2. Fine morning: wind NE, with *Cirrostratus,* which gave place to *Cumulus:* the evening was overcast, as for
rain, but little or none fell, and in the night there was a most copious
fall of dew.

RESULTS.

Winds Westerly, save twice about the last quarter of the moon, when
they became Easterly.

Barometer: Greatest height.............. 30·05 in.
Least...... .... 28·90 in.
Mean of the period........... 29·63 in.
Thermometer: Greatest height. .. ..,.... 75°
Least.....................,.... 34°
Mean of the period........... 57·65°
Mean of the hygrometer .... 52·3°
Rain....... .... 2·13 in.

In the present, as in the last Lunar period, the reader will observe
that the Last quarter, both in going off, and in approaching again, is
dry; while the other three phases have the rain divided among them.

Meteor.

" *Tunbridge Wells*, 6th *August*.—Being out about midnight, and the sky being remarkably clear, wind SW, therm. 49°, I saw in the WSW a brilliant meteor, almost half the apparant bigness of the moon: it began at about 45° or 50° of altitude, and slowly descended, increasing in size: it might perhaps be near 10 seconds in falling: the colour of the flame was white till near its extinction, when it was bright blue, tinged with reddish at the top. The day previous had been fair, with regular Cumuli, evaporating in the evening, and scarcely any other modification discernible all day. The day following had Cirrus passing to Cirrocumulus all the morning, with a south wind. T. F."

Atmospheric Electricity.

On the 16th of the Eighth Month, being at Tunbridge Wells, I raised a pretty large kite, made of linen stretched upon two pieces of cane, and fitted with a separate conducting string, as described vol. 1. p. xxx.; which being kept up at different elevations, for several hours in the afternoon, the following phenomena were remarked.

About noon, with a moderate breeze at SW, while the sky was clear, or only covered by light Cirrus passing to Cirrocumulus and Cirrostratus, the apparatus gave only weak and moderate sparks, which were of positive electricity. But when, after this, some obscurity had begun to appear in the sky to windward (the wind veering to S), and the rudiments of Cumuli, formed in a lower region, passed our zenith, the very first of these slight masses occasioned sparks attended with a sensation that reached to the elbows: a larger body of the same cloud soon after arriving, we received shocks extending quite to the feet, and too severe to be repeated often. At this time my son perceived a hissing noise, which seemed to attend the flying off of electricity from the reel: the sparks also became visible in full daylight, and a few drops of rain, from the skirt of a shower forming to windward, did not at all abate the charge. It was observed however by Dr. Forster, who was left with the apparatus between two and three p. m. that a cold breeze, which preceded a cloud bringing a shower, entirely took off the charge for a time. At three p. m. the shower becoming too heavy to permit us to continue the experiment satisfactorily, the *bearing* string of the kite was lowered, until the *conducting* thread touched the ground, and in this situation of the two strings the kite was taken in without our sustaining the repeated shocks which would probably have been encountered in that operation in the common mode of raising it with a single string.

During the experiment the barometer was falling from about 29·40 to 29·30 in.: and the temperature at the close was 58°. The sky, after the shower, was obscured by a double veil of Cirrostratus, and there fell more rain about half-past five p. m. We observed no negative charge on this occasion.

It is remarkable that the *Leyden phial* cannot be charged, beyond a certain weak degree of intensity, at the string of a kite : the charge is acquired by a single spark, and the person holding the phial is shocked by that and by each succeeding contact. It seems that the electric fluid passes (if I may be allowed the expression) with a *greater momentum*, when it has to glide down from a long elevated conductor, than when it is received by approach to a body insulated near the surface of the earth. I am aware, however, that the principles of electricity may supply a different explanation of this singular fact.

1817.		Wind.	Pressure. Max.	Pressure. Min.	Temp. Max	Temp. Min.	Hygr. at 9 a.m.	Rain, &c.
9th Mo.	L. Q. Sept. 3	SE	29·96	29·79	75°	53°	65	
	4	Var.	30·10	29·96	69	43	63	
	5	Var.	30·10	30·03	73	46	60	
	6	Var.	30·03	30·00	63	47	59	
	7	SE	30·03	29·91	71	44	60	
	8	E	30·07	29·94	76	47	58	
	9	NE	29·96	29·94	66	54	63	
	10	N	29·97	29·92	65	50	59	
New M.	11	NE	29·92	29·87	68	49	65	
	12	Var.	29·93	29·80	68	49	62	—
	13	SE	29·93	29·85	66	53	56	
	14	NE	29·93	29·83	63	56	63	0·17
	15	NE	29·98	29·93	66	60	64	
	16	SE	29·98	29·94	72	51	63	
1st Q.	17	NE	29·94	29·61	70	54	60	
	18	N	29·76	29·56	62	55	62	4
	19	W	29·95	29·76	67	47	57	
	20	NW	29·95	29·91	64	55	53	
	21	NE	29·91	29·83	63	42	61	
	22	NE	29·83	29·80	59	47	53	
	23	NE	29·90	29·80	64	47	52	
	24	SE	29·90	29·52	66	47	51	
Full M.	25	SW	29·52	29·16	64	55	65	—
	26	SW	29·31	29·16	60	47	58	0·20
	27	SW	29·65	29·31	58	44	48	7
	28	W	29·95	29·65	59	33	55	
	29	NW	29·95	29·89	58	43	54	
	30	NE	29·89	29·75	55	42	53	
10th Mo.	Oct. 1	NW	30·02	29·75	56	30	57	
	2	W	30·05	30·02	46	24	48	
			30·10	29·16	76	24	58	0.43

NOTES.—Ninth Mo. 3. Much dew: very fine day, with *Cirrus* only, in horizontal striæ: temp. 72° after sun-set. 4. Dew: fine morning: *Cirrocumulus*, followed by cloudiness from S, about nine: clear afterwards, save a line of low thunder clouds in the NE. 5. Fine, after misty morning: large *Cumuli*: at night the floating dust and smoke assumed the horizontal arrangement usual before the *Stratus*. 6. Misty morning: afterwards large plumose *Cirri*, passing to *Cirrocumulus*: p. m. some delicate streaks of *Cirrostratus*, with two currents near the earth at sun-set, SW above E. 7. Serene day, after misty morning : a very luminous, yellowish, evening twilight, with crimson streaks of *Cirrocumulus*, and a dewy haze round the horizon.   8. As yesterday, with *Cirri*, finely tinted in orange at sun-set.   9. Overcast, a. m. : at sun-set, *Cirrostrati* from SE.   10. Overcast morning : then *Cumuli*, with an electrical character : a fine breeze these three days. 11. Calm misty morning : then lightly clouded till evening.   12. Misty morning : after a little rain, the sky exhibited a veil of clouds moving from the SW.   13. *Cumulostratus* through the day.   14. Rain very early : temp. 63° at nine, a. m. : mild and damp air.   15. Cloudy, close, damp, day and night.   16. Overcast, with a breeze.   17. Misty morning : then sunshine and flying clouds.   18. Slight showers, with wind.   19. Cloudy morning : luminous evening twilight, orange, with rose colour above.   20. Clear dewy morning : the temp. scarcely varied from 55° through the night : *Cumulus*.   21—23. Fine, with breeze pretty strong, and various clouds.   24. The lateral approach from the southward of the westerly current was indicated to-day by the southing of the wind, by heavy *Cumuli* and *Cumulostrati* in the SE, and by a lurid haze, with greenish streaks of *Cirrostratus*, before the moon. 25. A gale from SW, with light rain : in the evening a lunar corona with *Nimbus* : heavier showers in the night.   26. Showery morning : then *Cumuli*, carried in a fine blue sky : evening showery : night windy.   27. Wind and showers.   28. The morning gradually cleared up, with *Cirrostratus* passing to *Cirrocumulus*, and some very elevated *Cirri* : at sun-set these showed red, stretching SW and NE.   29, 30. The wind, after going to SW for a short time, came round by N to NE, with fine weather.

Tenth Mo.   1 Fine : very red *Cirri* at sun-set.   2. Hoar frost, with ice.

RESULTS.

Winds Easterly, interrupted after the full moon by a gale from the westward.

Barometer : Greatest height .... 30·10 in.
Least .... 29·16 in.
Mean of the period .... 29·842 in.
Thermometer : Greatest height. .... 76°
Least .... 24°
Mean of the period .... 55·76°
Mean of the hygrometer .... 58°
Rain .... 0·48 in.

We have here, for the third time in the space of three months, a striking relation between the occurrence of rain, and the times of the Moon's being in opposition and conjunction : but in the present case the rain followed instead of preceding or accompanying those phases. The Last quarter is dry, as in the two periods before, and the First nearly so.

1817.		Wind.	Pressure. Max.	Min.	Temp. Max.	Min.	Hygr. at 9a.m.	Rain, &c.
10th Mo. L. Q.	Oct. 3	NE	30·14	30·05	50	32	65	
	4	NE	30·22	30·14	57	32	55	
	5	NE	30·25	30·22	56	33	56	
	6	NE	30·25	30·10	54	39	55	
	7	E	30·10	30·06	54	43	51	
	8	NE	30·06	29·95	54	35	46	
	9	NE	29·95	29·87	53	39	59	
New M.	10	N	29·92	29·86	56	41	56	
	11	NE	29·99	29·92	50	36	54	—
	12	N	30·15	29·99	52	35	55	5
	13	NE	30·22	30·17	52	32	59	—
	14	N	30·22	30·02	50	42	64	—
	15	NW	30·02	29·80	48	37	64	6
	16	NE	30·05	29·80	48	36	60	30
1st Q.	17	NE	30·05	29·86	48	37	59	5
	18	E	29·87	29·77	45	37	54	17
	19	N	29·91	29·87	45	42	62	
	20	N	29·89	29·85	48	40	64	3
	21	E	29·79	29·77	52	39	65	
	22	N	29·90	29·79	48	36	65	
	23	NE	29·90	29·88	50	40	63	—
	24	NE	29·88	29·81	46	38	65	4
Full M.	25	SE	29·81	29·71	50	37	65	
	26	S	29·69	29·65	52	28		
	27	SW	29·43	29·32	49	32		12
	28	SW	29·41	29·39	48	32		16
	29	SW	29·55	29·21	49	27		8
	30	S	29·49	29·14	57	42		21
	31	S	30·25	29·46	52	28		7
11th Mo.	Nov. 1	Var.	30·34	30·16	49	27		
			30·34	29·14	57	27		1·34

Tenth Mo. 3. Hoar frost, with ice. 4. A strong breeze: clear morning: the wind, p. m. tending to SE, with *Cumuli.* 5. The same breeze still: *Cumulus,* succeeded by *Cumulostratus,* which became heavy by noon; when the smoke of the city, being drawn up in a column in the SW, mingled with the clouds, and gave occasion (as it appeared) to a local shower: it drizzled a little with us, and there was a bank of clouds beneath dewy haze in the NE at sun-set. 6, 7. Some wind, especially by night: *Cumulostratus.* 8. *Cumulus,* &c.: windy. 9. *Cumulostratus:* windy: SE, p. m. 10. *Cumulostratus,* somewhat heavy, with an excessive rising of the dust in the evening. 11. A fresh breeze again, with fleecy *Cumulus* and *Cirrostratus:* a little rain, mid-day: very fine orange twilight. 12. As yesterday: slight showers by inosculation of different clouds: the product of the rain-guage includes much dew. 13. A strong breeze, NNE, a. m.: some drizzling rain: twilight fainter orange. 14. Misty morning: a little drizzling, p. m. 15. Some rain, mid-day and evening. 16. Wind got back to NE, and fresh, a. m.: rainbow at eleven: wet, mid-day: then cloudy. 17. Showers, with hail, about noon. 18. Cloudy: wind fresh, going first to NW, then back to E: showers. 19. Temp. 45° at nine, a. m.: dark and cloudy through the day. 20. Gloomy: misty: but little wind. 21. The same. 22. Lighter sky: wind to SE, and at night back to N. 23. Wind brisk at NNE, with a lofty sky: a shower, p. m. 24. Drizzling: dark, a. m. 26. Fair: *Cumulus*, with *Cirrocumulus* and *Cirrostratus:* the moon rose gold-coloured: a clear night ensued. 27. Misty morning: rain, with wind, at night. 28. Misty: the trees dripping: the wind to S, then back to SW, with pretty heavy rain. 29, 31. After a moderate gale from the southward, the barometer rose rapidly, with squalls of wind, showers, and hoar frost.

Eleventh Mo. 1. Very fine day: misty at night, probably from a *Stratus*.

RESULTS.

The wind, which was chiefly from the NE to the time of the Full moon, came round afterwards (as in the last period) to the SW for a few days only.

Barometer:	Greatest height....	30·34 in.
	Least	29·14 in.
	Mean of the period..	29·881 in.
Thermometer:	Greatest height.	57°
	Least	27°
	Mean of the period..	43·27°
	Rain.....	1·34 in.

The hygrometer having been out of order, the latter week's observations on it are uncertain. I found the *evaporation* to proceed of late in the following ratio, viz. :—

In eight days preceding the 3d of 10th month.	0·42 in.
In seven days preceding the 10th...	0·37
In seven days preceding the 17th.	0·22
In eight days preceding the 25th....... .	0·10
In eight days to the close of the period (with considerably more wind stirring).	0·12

The capacity of the air for water has, therefore, decreased more rapidly than the daily mean temperature, the approaching change of the atmospheric current being the probable cause.

HURRICANE IN THE WEST INDIES.

A Hurricane raged in the West India islands on the 21st of the Tenth month, which is thought to have been more destructive than any since the year 1780. Its ravages appear to have extended in breadth at least from the 12th to the 18th degree of north latitude, but to have affected principally the islands of Dominica, Martinique, St. Lucia, and St. Vincent; and, in a less degree, Barbadoes, which lay to leeward with regard to the storm. The wind is stated to have set in at day-break from the NW and to have raged with tremendous violence, with occasional falls of rain, until three in the afternoon, when becoming southerly it abated, but did not immediately cease.

At St. Lucia, the government house and barracks were blown down and their inmates buried in the ruins; the crops and forests, in short the whole face of the island, desolated, and every vessel in the port lost. Some of the numerous documents inserted in the papers relative to this disaster may be suitably presented to the reader as descriptive of these tremendous visitations.

OFFICIAL.

The following Letter on this lamentable occasion was addressed by the Colonial Secretary of St. Lucia to the Governor of Barbadoes:—

" *St. Lucia, Oct.* 23, 1817.

" My Lord, — His Excellency Major General Seymour being unable to address your Lordship, in consequence of the very serious injury he received during the hurricane of the 21st instant, (and, I am sorry to add, very little hopes are entertained of his recovery), his Excellency has directed me to give your Lordship the particulars, and earnestly to entreat your Lordship's assistance and support towards ameliorating the situation of the unfortunate inhabitants of this island, not one of whom but has suffered severely thereby.

" Scarcely a dwelling or a negro house is left standing : the mills and outbuildings either unroofed or razed to the ground ; nearly the whole crop of canes torn up by the roots, and the face of the island, which was luxuriant on the 20th, now bears the appearance of an European winter.

" The town of Castrees is nearly in ruins, and the vessels, about 12 sail, are on shore, not one of which is expected to be saved. The whole of the buildings of Morne Fortunée and Pigeon Island were blown down, with the exception of the magazine and tanks.

" His Excellency and family were taken from under the ruins of his residence (the Commandant's quarters), where he remained in the hope that it would have resisted the gale ; but he has, unfortunately, suffered for his imprudence.

" I have the honour to be, my Lord, your Lordship's most obedient Servant,

" L. R. Baines, Colonial Secretary."

" P.S. Since writing the above I am sorry to acquaint your Lordship, that great fears are entertained that General Seymour cannot pass 48 hours."

Extract of a Letter from St. Pierre, Martinique, dated November 10, 1817.—" On the 21st of October, this Colony was visited by the most furious hurricane ever witnessed here; the details of this sad disaster would be equally long, as painful. The loss of nearly 1800 lives, 25,000 hhds. of sugar of the present and next crop, incalculable losses in buildings, animals, and the necessaries of life, have occasioned a general desolation, independently of the great anxiety caused by nine-tenths of the shipping, which were in different ports of this island, being either wrecked, damaged, or missing. St. Lucia and Dominica have equally suffered ; the tempest reached also St. Vincent's and Grenada. Its ravages also extended to Guadaloupe, as well as Porto Rico and its neighbourhood, though in a less degree. The loss experienced by Martinique alone, may be very moderately calculated at 25,000,000 of francs, exclusive of the shipping. The works and buildings of entire parishes were razed to the ground; it lasted 26 hours, 12 of which with such inconceivable fury as to produce all these disasters, and to destroy buildings which had withstood all former hurricanes. — It will require many years before the colony can recover itself from this heavy calamity."

A Letter from an Officer on board His Majesty's ship Antelope, at St. Kitt's, dated Nov. 12, says: —" We were lying at St. Lucie quietly at anchor, only the day before the hurricane came on, and got under sail for Barbadoes (as was our intention) about seven in the morning of the 20th, it being fine weather. At twelve o'clock the night following the officer of the watch hailed the master, and said the wind had come round to the north-west, which was very unusual in this country, where easterly winds prevail all the year through, and that the weather appeared to be coming on bad; the Admiral and Captain were immediately upon deck; we took in all our sails except the fore sail, which was reefed; got the top-gallant-masts upon deck, and prepared for the worst. At three in the morning of the 21st instant, it blew very hard at west, with tremendous heavy rain; at four still harder: took in our fore-sail, and brought her to under a try-sail double-reefed; at seven yet harder, when a sea came and washed away one of our boats from the stern. When ten o'clock came it blew a perfect hurricane beyond what any of us had ever witnessed; however, the ship lay very quiet and behaved very well, but from the heaviness and quantity of rain, with the immense force of wind, all our cabins were full of water. At noon the weather abated, and at three in the afternoon it became quite moderate, when we set our sails again.—We visited St. Lucie sixteen days afterwards, and the scene was such as my pen cannot describe: many of the inhabitants lost their clothes, and those who are sick, are lying on the ground with no other covering than the sky, exposed to sun and rain.

Extract of a Letter from an Officer on board one of His Majesty's ships, dated *Barbadoes, November 30, 1817:*—

" On our first making the Island (St. Lucie) we were struck with astonishment at the total change in the whole face of the country. We left it the day before the hurricane a beautiful rich green, and every thing in a most flourishing state. It has now the appearance of a severe European winter. We went on shore on the 7th of November; the scene of destruction which then presented itself is far beyond my power of description. On Pigeon Island, three houses only are left standing out of nearly 250, the rest, with the church, are almost totally demolished; one of the three is shifted 15 feet off its foundation, without going to pieces. The two large tamarind trees, under which the Negroes always met to dance on Saturday and Sunday evenings, were torn up by the roots, which by their spread in this rather sandy soil, nearly equal the branches in circumference. The woods with which this island particularly abounds are more or less scattered, according to their exposure to the gale; many of the trees which are left standing, have only a few of the ragged stout branches remaining. In the deep ravines the wind appears to have acted in a whirl; for immense trees are completely thrown down and twisted up in heaps in a most astonishing manner. The inhabitants tell us the great hurricane in 1780 was not equal to this."—*Pub. Ledger.*

From the India Gazette.

Bombay, *October* 20, 1817.—Until yesterday the rain has continued, during the last week, to come down almost without intermission: the quantity which has fallen in this monsoon is uncommonly great, the accounts taken by different rain guages give upwards of 95 inches since the 2d June, being above a fourth more than the average fall of former years. The weather has now cleared up, with every appearance of continuing fair and hot."

The following is an Extract of a Letter from St. Petersburgh.

" Advices from Georgia of the 13th ult. state, that on the 21st of October, an immense avalanche fell from the mountain of Kasbeck, and covered an extent of three wersts in length, to the height of 50 fathoms. It completely dammed up the rapid river of Tereck; which, however, on the third day, worked a passage for itself underneath the mass of snow. This accident for some time interrupted the communication with Georgia. Fortunately, there were no travellers passing when it happened. Nine years have elapsed since the last avalanche occurred: though, according to the reports of the inhabitants, they generally take place once in seven years, and in the summer."—*Pub. Ledger, Jan.* 14.

TABLE CXXXVII.

1817.		Wind.	Pressure. Max.	Pressure. Min.	Temp. Max.	Temp. Min.	Hygr. at 9 a.m.	Rain, &c.
11th Mo. L. Q. Nov.	2	S	30·19	30·11	57°	41°		
	3	SE	30·10	30·06	54	49		
	4	SE	30·06	29·91	55	45		
	5	SE	29·88	29·85	54	44		
	6	S	29·85	29·69	55	49		6
	7	SW	29·69	29·43	58	50		13
New M.	8	SW	29·61	29·35	55	44		5
	9	W	29·90	29·61	51	40	63	
	10	SW	29·89	29·66	55	44	95	
	11	SE	29·70	29·68	52	38	62	14
	12	SW	29·73	29·49	55	38	100	2
	13	SE	28·73	29·48	52	38	70	—
	14	SE	29·48	29·26	54	49	86	76
1st Q.	15	W	29·79	29·26	55	40	80	10
	16	SW	30·15	29·79	54	39		
	17	SW	30·17	30·16	59	50	83	
	18	W	30·38	30·16	57	35	67	
	19	NW	30·45	30·43	49	30	85	
	20	SW	30·43	29·87	46	38	96	
	21	N	30·12	29·87	52	37	72	
	22	NW	30·12	30·05	47	37	73	
Full M.	23	W	30·05	29·82	46	41	65	
	24	W	29·78	29·70	49	32	71	—
	25	W	30·00	29·78	44	34	64	15
	26	NW	30·07	30·00	52	40	78	
	27	SW	30·08	30·06	51	37	65	
	28	SW	30·06	29·92	49	40	67	
	29	SW	29·93	29·87	54	49	78	3
	30	SW	29·87	29·79	54	52	79	5
12th Mo. L. Q. Dec.	1	SW	29·79	29·54	54	46	80	51
			30.45	29·26	59	30	76	2·00

NOTES.—Eleventh Mo. 2 to 4. Nearly calm: dripping mists, with alternate obscurity by *Cirrostratus*, and sunshine. 5. A slight shower at night. 7. A gale at SW: showers by night. 8. Squally: several showers in the day. 9. Windy, fine, with *Cumulus, Cirrus*, and *Cirrostratus.* 10. Misty morning: *Cirrus, Cirrostratus*: fair day: windy at night. 11. Fine morning: then quickly overcast and wet, a. m.. 12. Red sun-rise: then *Cirrostratus*, speedily general: cloudy till evening: windy at night. 13. Fine morning, with *Cirrostratus*: fair day: rather windy night. 14. Much *Cirrostratus* in the morning:

rain by half past nine: fair evening. 15. Wet, gloomy morning: calm and lighter, mid-day: p.m. the wind went to W, and blew strong: clear night. 16. Coloured *Cirri* with *Cirrostratus* at sun-rise: misty: steady breeze, with some appearance of distant rain: cloudy evening. 17. Fair: somewhat windy night. 18. Fair: *snow* fell within a few miles of us: evening twilight luminous and orange-coloured. 19. *Cirrocumulus, Cirrus* and *Cirrostratus*: abundant dew on the grass all day: very fine sky. 20. Very misty, a.m.: the trees drip much: fine, p.m. with dew and large *Cirri*. 21. Cloudy: rather windy: little or no dew this morning: *Cirrostratus, Cumulus*: the wind got to N at night. 22. Fair: *Cirri* in lofty bars, stretching N and S, followed by *Cirrocumulus*, and a group of clouds among the smoke of the city. 23. Fine, clear morning: grey sky after. 24. *Cirrostratus* with *Cirrus*, at sun-rise: a little light rain, p.m.: lunar corona, followed by a large faint halo. 25. Hoar frost: a steady gale through the day, with an appearance of *Nimbi* in the NW: rain after sun-set. 26—28. Fair: somewhat windy, with *Cirrostratus*, &c. 29. The hygrometer stood at 78° till noon: a little rain fell, a.m. and at night: the bees came out in considerable numbers, continuing however about the hive. 30. Overcast, windy: the maximum temp. at nine, a.m. or rather, the whole 24 hours warm alike.

Twelfth Mo. 1. A wet day.

RESULTS.

Winds Southerly and Westerly.

Barometer: Greatest height .... 30·45 in.
Least .... 29·26 in.
Mean of the period .... 29·878 in.
Thermometer: Greatest height .... 59°
Least.... .... 30°
Mean of the period.......... 47·11°
Mean of the Hygrometer ... . 76°
Rain...................... 2 in.

Evaporation, 1·02 inch, divided as follows: to the 9th, 0·27; to the 15th, 0·40; to the 23d, 0·23; to the end, 0·12.

The *mean temperature* of this period instead of falling some degrees (as might have been expected from the season) is near 4° higher than that of the last. But in the Tenth month it had undergone a disproportionate depression of 12½ degrees, (as will appear on comparing together the results of Tab. 135 and 136) for which the present warmth may be considered as a compensation.

The Rain appears now to be quitting the vicinity of the times of New and Full Moon, and attaching itself to the quarters. Yet the space of three or four days following the Last Quarter is dry, as it has been in four successive periods since the Summer solstice. It will be found however, on examining the Tables, both forward and backward, that this portion of the Lunar period is subject at times to heavy rain.

E

1817.		Wind	Pressure.		Temp.		Hygr.	Rain,
			Max.	Min.	Max.	Min.	at 9 a.m.	&c.
12th Mo. L. Q.	Dec. 2	NW	29·45	29·43	47	29	71	2
	3	NE	29·86	29·40	41	28	78	5
	4	E	29·87	29·71	42	29	83	
	5	S	29·87	29·37	47	38	69	27
	6	NW	29·38	29 37	41	28	83	
	7	SW	29·37	28·61	42	32	92	13
New M.	8	Var.	29 07	28·54	45	36	85	48
	9	NW	29 22	29 07	43	28	69	
	10	NE	29·51	29 22	35	21	73	—
	11	Var.	29·67	29·51	32	18	80	
	12	SE	29·67	29 61	37	23	83	—
	13	SE	29 61	29·48	40	35	85	40
	14	SE	29·57	29·34	48	32	98	24
1st Q.	15	SW	29·80	29·57	43	34	97	72
	16	SW	29·53	29·30	50	35	96	32
	17	SW	29·53	28·88	48	35	74	67
	18	SW	28·88	28·74		34	65	2
	19	Var.	29·30	28·88	46	39	76	15
	20	NE	29·56	29 50	40	27		
	21	NE	29 50	29·44	33	28		
	22	N	29·53	29 46	33	29	74	
Full M.	23	SE	29·65	29·48	32	20	65	
	24	NE	29·90	29·65	32	24	80	
	25	NE	30 06	29·90	33	22	78	
	26	Var.	30·08	29·80	36	22	77	10
	27	SW	29·62	29·45	37	32	92	5
	28	NW	30·10	29·62	38	21	72	
	29	SW	30·10	29·97	33	23	65	
	30	Var.	30·00	29 93	40	22		6
L. Q.	31	NE	30·01	30·00	35	21	93	
			30·10	28.54	48	18	78	3·68

NOTES.—Twelfth Month. 2. Some rain, a. m. 3. Hoar frost; *Cirrostratus*; the sky quickly overcast: some sudden showers followed: in the evening *Nimbi*, the wind going to NE with force. 4. Hoar frost: fine, with *Cirrostratus*: misty p. m.: windy night. 5. Very cloudy, a. m.: a gale in the night, followed by rain. 6. Overcast, a. m.: fine and calm mid-day: *Cirrocumulus*. 7. Hoar frost: *Cirrostratus*, with a stormy appearance, a. m.: fine, p. m.: rain, with a gale of wind, in the night. 8. *Cirrostratus* at sun-rise, the wind gone down: wet at intervals: *Nimbi*: much wind again at night. 9. Cloudy, a. m.: about noon *Cirrostratus*, and after it *Cirri* in elevated bars stretching N and S, coloured red at sun-set : starlight, with small meteors. 10. Snowing by nine, a. m.: at eleven the ground was white, when it ceased: clear night. 11. The sun emerged from *Cirrostrati:* rather misty air: the wind gentle at SW, a. m.: but easterly in the night. 12, a. m. Vane at SW: calm: much rime, with a misty air till evening : a thaw in the night. 13 Obscure by *Cirrostratus*, a. m.: the hygrometer proceeding towards moisture : rain, gentle in the day, heavier, with wind, in the night. 14. *Cirrostratus* prevailed in a uniform close canopy about the height of the neighbouring hills, on which I found it misty in consequence, while small rain fell below : early in the night came on wind, with showers. A perfect, but colourless *lunar bow*, was observed about ten, p. m. and reported to me by a gentleman whom I met at Stamford Hill in the morning. 15. Hoar frost: the sun emerged from a low *Cirrostratus:* very wet, p. m. and night. 16. Wet, a. m.: in the night a heavy gale, ceasing about three. 17. Fair, a. m.: obscure afterwards by *Cirrostratus :* in the night a most violent westerly gale, increasing and decreasing in force by slow intervals, with much rain. 18. Windy: bright moonlight. 19. Wet, p. m. 20. The wind got to NE, a fresh breeze, but at night the clouds came from NW. There was a manifest attraction between the low clouds and the smoke of London. 21. Fair: the clouds tending to *Cumulostratus*. 22. A very slight sprinkling of snow, crystallized in stars. 25. The same, in grains as fine as basket salt. 26. Orange-coloured sun-rise, with red *Cirri*: hoar frost : a lunar corona last night, surrounded by a coloured halo. 27. After fine dry frost for some days, a thaw early this morning, with rain : in the night a gale, with showers, after which a ground frost. 28. A little snow at mid-day: the temp. 19° on the ground at night. 30. Wet, p. m. 31. A frozen mist came on at eight, a. m. from the southward ; and after a clearer interval there was again a very thick fog in the evening.

RESULTS.

Winds variable.

Barometer:	Greatest height	30·10 in.
	Least	28·54 in.
	Mean of the period	29·508 in.
Thermometer:	Greatest height	48°
	Least	18°
	Mean of the period	32·66
Mean of the hygrometer		78°
	Rain	3·68 in.
	Evaporation	0·38 in.

LARGE METEOR.

" On the morning of the 8th inst." (Dec.) says the observer, who dates from·
Ipswich, " I was looking at Mars, whose position is near to the star in the bull's
northern horn. About midway between the horns I suddenly perceived a fiery body
resembling a red-hot ball of iron, four or five inches in diameter, which having
passed three or four degrees in a direction between the principal stars of Capella
and Canis minor, burst into a spherical body of white light nearly as large as the
full moon, of so great lustre as scarcely to be borne by the eyes, throwing out a
tail about three degrees in length of a beautiful rose colour tinged round the edges
with blue. It thus proceeded in its course without apparent diminution towards
the principal star in the head of Hydra, (very near to the ecliptic,) a little beyond
which it suddenly disappeared (I believe) with an explosion; as I distinctly heard
a rumbling noise like that of cannon discharged at a distance, about ten or twelve
seconds afterwards. Its duration as nearly as I can estimate was about five seconds,
during which it traversed a space of nearly sixty degrees. It is scarcely possible
to give an adequate description of the vivid splendour which characterized this
extraordinary phænomenon. It cast a light around equal to the noon day's sun :
I could compare it to nothing so well as the beautiful dazzling light exhibited by
the combustion of phosphorus in oxygen gas; its effect upon the organs of light
being analogous. The barometer was falling at the time, and in the course of the
night fell altogether an inch and one tenth; the thermometer at 42° Within a
quarter of an hour afterwards the atmosphere became entirely obscured by clouds;
violent tempests of wind and rain succeeding, although the stars were previously
visible and the zenith free from vapours." J. A.— *Philo. Magazine.*

The most remarkable circumstance in this account is, that the eye
of the spectator having been favourably directed at the moment, he
had an opportunity of seeing (what I believe is very seldom witnessed)
the *beginning incandescence* of one of these bodies, and of tracing its
motion before it had become completely ignited. Leaving for a future
occasion any general reasoning which might be founded on this, I may
observe that as these meteors are not *necessarily* very elevated or very
distant, the appearance of the present one *may* have had some connection with the low Barometer, the supervening cold current, and
the copious precipitation of water, which we find registered upon and
subsequently to the day of its appearance. It may be remarked also
that *small meteors* were sufficiently frequent on the following evening,
to induce me to note their occurrence.

GALES OF WIND AND THUNDER STORMS.

The storm which raged in the night between the 7th and 8th of December,
destroyed in the Channel 20 vessels from the coast of Brest to St. Maloes.

Falmouth, Dec. 8.—It has blown all day a hurricane, from W to NW, and still
continues with unabated violence. The gale commenced about 12 o'clock last
night, but up to this time (9 a.m.) the East and West Indiamen ride it out, and no
damage of any kind has yet happened.

A dreadful storm was felt in the Bay of Biscay on the 9th, 10th, and 11th December; a vessel was wrecked on the bar of Bayonne; and the plains of the Adour and the Nive were inundated. At St. Jean de Luz, the dyke which defended the town from the ocean was much damaged, and the inhabitants seriously alarmed.

We fear we shall receive more distressing accounts of the damage done on our coasts by the severe gales which have blown since the beginning of the present month. Last Sunday night, and on Monday (Dec. 14—15), there was another storm; and the agent to Lloyd's, at *Poole*, represents it as most tremendous. It was accompanied with the most awful thunder and lightning.—*Pub. Ledger.*

A *Whitehaven* paper says — The weather in the course of the last week, has proved the most tempestuous that has been experienced in this part of the country for a long time past.—On the night of Monday, the 15th Dec. we had one of the most tremendous thunder storms perhaps ever known in this part of the kingdom. Its greatest violence was between the hours of nine and ten. The peals were awfully loud, and the lightning astonishingly vivid and frequent.

The *Hull Advertiser* of the 20th Dec, says — On the nights of the 16th and 17th, this town and neighbourhood were visited by a most violent gale of wind attended with heavy rain, of the effects of which in some parts of the country the accounts are very distressing.

Jersey, Dec. 19.— For some days past, perpetual storms, attended with heavy rain, mixed with hail, have prevailed here, wind varying from N to NW W and SW.

CITY FOGS.

The fog of Wednesday (Dec. 31,) seems to have been confined to the Metropolis and the immediate vicinity. No further to the Northward than the back of Euston-square, the weather was clear and even bright. A gentleman, who came to town from Enfield, saw no fog till he approached London. Southward of London, it extended as far as Clapham, and it was rather thicker in some of the environs, than in the Metropolis itself. Upon an average, ten feet was the distance, at which objects became invisible, out of doors. Within doors it was impossible to read without a candle.

The following paragraph from *The Freeman's Journal* of January 1, shows, that the fog on Wednesday last was as thick in Dublin as in London:—'The oldest person living has no recollection of a fog so thick as the one which enveloped this city last evening, between the hours of six and nine. It was more dense in some streets than in others, and where this was the case it was impossible to pass with convenience without the aid of opened lanterns.—*Pub. Ledger.*

	1818.		Wind.	Pressure.		Temp.		Hygr. at9a.m.	Rain, &c.
				Max.	Min.	Max.	Min.		
1st Mo.	Jan.	1		30·01	29·96	33°	21°	90	
		2	SE	29·96	29·58	33	31	80	
		3	SE	29·58	29·43	40	32	75	13
		4	S	29·64	29·48	44	32	100	—
		5	Var.	29·90	29·48	42	28	80	51
New M.		6	SW	30·15	30·10	41	29	85	
		7	SW	30 03	29·82	47	36		17
		8	NW	30·18	30·05	41	29	58	
		9	SW	29·85	29·76	48	39	83	11
		10	SW	29·75	29·72	52	47	93	3
		11	NW	29·76	29·50	49	36	80	9
		12	SW	29·92	29·78	47	37	60	3
		13	SW	29·85	29·79	50	40	77	—
1st Q.		14	SW	29·83	29·60	50	39	95	24
		15	W	29 80	29·56	53	45	70	—
		16	W	29·76	29·46	50	35	58	16
		17	W	29·88	29·85	40	32	60	
		18	Var.	30·40	29·85	42	28	65	—
		19	SW	30·43	30·30	38	27	70	
		20	Var.	30·30	29·85	40	28	95	
		21	Var.	30·12	29·92	45	26	89	
Full M.		22	SW	29·65	29·40	45	31	90	12
		23	Var.	29·60	29·42	41	28	72	18
		24	NW	29·93	29·39	43	29	93	9
		25	W	28·87	29·79	47	34	72	1
		26	SW	29·88	29·60	52	32	76	1
		27	SW	29·88	29·50	46	35	72	17
		28	NW	29.65	29·50	43	28	85	—
L. Q.		29	SW	29·65	28·98	43	32	77	27
				30·43	28·98	53	21	78·5	2·32

NOTES.—First Mo. 1. Fine. 2. Much wind in the night: snow by morning. 3. After some snow, a thaw, p. m. 4. Small rain, a. m.: fair, p. m.: starlight. 5. Wet stormy day: clear night. 6. Hoar frost: fair. 7. Fair, a. m.: cloudy, p. m.: a gale, with rain, during the night. 9. Wet mid-day: the *Cirrocumulus* has appeared two or three times within a week past. 10. *Cirrus* and *Cirrostratus* at sun-set, rose-coloured, followed by a gale with rain. 11. Gloomy, a. m. with small rain, p. m. and night as yesterday. 12. Fine day: a gale again in the night. 13. The gale continued, with cloudy weather:

the night, after a calm evening, was stormy. 14. Wet stormy day: in the evening, *Cumulus*, with *Cirrostratus* and *Cirrus*: the air clearing: much wind in the night. 15. Much cloud and wind, a. m.: small rain, p. m. night stormy. 16. Fair, with a gloomy sky, a. m.: some rain, p. m.: a heavy gale in the night. 17. The wind is now more moderate, with a tendency to NW: a very fine day and night. 18. Fair, with a breeze: a squall, with a little rain, p. m.: bright moonlight. 19. Fine. 20. Red *Cirrostrati* at sun-rise, with hoar frost: fine day, with *Cirrostratus* in flocks: *Cirrocumulus*, and *Cirrus*, with a rainy aspect. 21. Gloomy overcast morning: some wind and rain by nine: afterwards fine with *Cumulus*, and a breeze from NW: lunar halo. 22. Fine: windy, p. m.: in the fore part of the night, a heavy southerly gale, with showers. 23. Fine, windy: the barometer fluctuating: lunar corona at night, the wind NW. 24. Rain very early: wet forenoon: fair p. m. and night windy. 25. Fine, a. m. with *Cirrus* and *Cirrostratus*: windy night. 27. Much wind in the night, followed by rain. 28. Large *Cumuli*, with *Cirri* above, and the rapid developement of *Cumulostrati*, presented this afternoon a spring-like sky. 29. Hoar frost: fair day: a thickness to the S and W p. m. was followed by a nocturnal gale, with rain, as usual of late.

RESULTS.

Winds Westerly, introduced by the South East.

Barometer: Greatest height .... 30·43 in.
　　　　　　Least......... .... 28·98 in.
　　　　　　Mean of the period .. 　　..... 29·786 in.

Thermometer: Greatest height .... 53°
　　　　　　　Least . 　.............. 21°
　　　　　　　Mean of the period..... ...... 38·46°

Mean of the hygrometer . .. . 78·5°
　　　　　Rain . ... 2·32 in.

Character of the period, stormy and changeable : amidst a succession of gales of wind, there were many intervals of fine weather by day ; and, as the thermometer shows, of frost by night.

Edinburgh, Jan. 3. — Notwithstanding the severe and continued frost, very little snow has hitherto fallen: excepting a little between Newcastle and Berwick, where it may be six inches deep in some places, there is none to be seen on the whole line of road from London.

Storms of Wind.

Edinburgh, Jan. 14. — On Monday night (12th) we were visited by one of the most severe gales we have experienced for a long time. It began to blow about ten o'clock from the south-west, accompanied with heavy rain, and continued to increase during the night until it became a perfect hurricane.

Edinburgh, Jan. 17.—Thursday morning (15th) the barometer had fallen eight-tenths of an inch; it then blew very hard, and during the whole course of the day slates and chimney-pots were flying about in all directions.—In the evening the gale increased, and about five o'clock it blew a perfect hurricane. In houses fronting the west a good deal of mischief was done in breaking the panes of glass, stripping the lead from the roof, dashing the cupola windows from their frames, and shivering them to atoms. In the course of the forenoon, two of the small minarets on the top of St. John's Chapel, at the west end of Prince's-street, gave way, and fell without doing any material damage to that beautiful building; not so, however, the effects of the evening—the violence of the wind carried off the whole of the minarets, large and small, leaving the summit of the tower a perfect ruin. Such was the force of the wind, that the masses of masonry were carried 30 feet beyond the base of the tower, penetrating not only the roof of the church, but also the floor, and breaking through the vaults to the foundation. One of the solid bars or bats of copper by which one of the pinnacles was bound to the top of the tower, was above an inch broad, and 5-6ths of an inch thick; and though the pinnacle to which it belongs was only six feet high, with a medium breadth of about eight inches, so as to expose a surface of merely four square feet, yet such was the power of the wind, that it tore the copper bar from its place, and twisted one of its arms, which was eight inches long, through an arch of 90 degrees, as if it had been a slender piece of lead. This effect resembles more that which is sometimes produced by lightning than by any other agent.

Large Meteor.

A beautiful meteor, with a long train, was observed at *Campbletown*, near Fort St. George, (Highlands,) at six o'clock on Wednesday (Jan. 28?). To the naked eye, the diameter of the ball appeared to be about one foot, and the length of the train about six feet. Its course was from West to East.—*Pub. Ledger.*

A violent storm was experienced at *Hamburgh* on the 15th Jan. accompanied by hail and rain. The Elbe rose so high, that all the lower part of the town was inundated, and the streets could only be passed in boats. Much damage was sustained, and melancholy accounts were expected from sea.

Feb. 5.—Two Hamburgh Mails arrived yesterday. From the Papers by them it appears that violent storms, accompanied with hail, thunder and lightning, have been experienced in most of the northern parts of the Continent. At Adensee, Stettin, Konigsberg, &c. much damage was sustained.

A Letter from *Carlisle*, dated *January* 24, says:—During the week the weather has varied much; boisterous, rainy, and frosty, in succession. The Eden, Petterill, and Caldew, were much swollen by the mountain torrents.—*Pub. Ledger.*

1818.		Wind	Pressure. Max.	Pressure. Min.	Temp. Max.	Temp. Min.	Hygr. at 9 a.m.	Rain, &c.
1st Mo.	Jan. 30	W	29·23	28·85	44°	33°	70	7
	31	SW	29·23	29·09	39	29	63	18
2d Mo.	Feb. 1	Var.	29·09	28·94	41	24	74	—
	2	NE	29·10	28·94	32	20	90	—
	3	NE	29·17	29·09	37	24	88	
	4	N	29·50	29·09	32	26	82	
New M.	5	Var.	29·90	29·50	—	—	88	
	6	Var.	30·02	29·90	41	24		
	7	Var.	30·02	30·00	31	23	92	
	8	NW	30·00	29·93	33	20	89	
	9	Var.	29·99	29·93	33	23	90	
	10	E	30·15	29·99	33	25		
	11	SW	30·16	30·15	37	29	75	
	12	SE	30·16	30·05	35	30	72	
1st Q.	13	NE	30·05	29·81	38	23	70	
	14	SE	29·84	29·72	36	25	90	
	15	E	29·88	29·80	40	27	93	
	16	NE	29·93	29·88	47	34	97	
	17	S	29·93	29·89	—	—	99	
	18	SW	29·93	29·80	49	34		23
	19	SE	29·99	29·75	51	28	95	19
	20	S	29·96	29·65	46	29	95	6
Full M.	21	SW	29·65	29·22	45	34	72	15
	22	NE	29·73	28·94	38	28	67	—
	23	NW	29·74	29·50	41	31	72	1·62
	24	SW	29·80	29·51	49	36	61	—
	25	SW	29·51	29·37	52	35	62	—
	26	NW	29·72	29·42	42	29	72	—
L Q.	27		29 60	29·27		34	82	57
	28		29·53	29·29	47	35	62	3
			30·16	28·85	52	20	80	3·10

NOTES.—First Mo. 30. Squally : showers, a. m. : wind and rain in the night.

Second Mo. 1. Hoar frost : rain, followed by snow in large flakes : *Cumulostratus* and *Nimbus*. 2. Hoar frost : fine *Cirri*, with *Cirrostratus*, a. m. : the lower modifications with some loose snow, p. m. 3. Fair : hoar frost. 4 Cloudy morning : snow on the wagons coming from the north, probably of last night : it seems to be the wind blowing over snow that keeps down our temperature to 32°. 5. A few drops of rain about nine, a. m. : after which very fine, with *Cumulus*, &c. and inosculation. 6. Hoar frost : misty. 7. Hoar frost : the paths icy: mist increases, with a calm air. 8. Misty: rime to the tops of the trees. 9. Misty: the rime falls partially. 10. Misty till evening, when it cleared up, and the rime fell off. 11—15. Hoar frosts, with fine weather: *Cumuli*, &c.: on the 15th a few drops of rain. 16. Very fine, with *Cumuli*, &c.: at seven, p. m. a large faint lunar halo. 17. Fine, with *Cirrocumulus*, &c. 18. Windy: wet, p.m. 19. A slight *Stratus*, a.m.: wet evening: windy night. 20. Hoar frost; somewhat misty; rain before noon: some hail in a shower at half-past one: after this *Cumulus*, with *Cirrus* and *Cirrostratus* above, the latter *imbricated*, or overlapping, like the branches of a pine-tree; then *Nimbi* amidst groups of other clouds, the lofty crowns of which were long coloured with a fine gradation of red tints about sun-set: the sky around the moon showed violet, while the disc was brassy. 21. Much wind, a.m. with clouds driving high and close; wet, p.m.: at evening a lighter sky, with *Cumulus* and *Cirrocumulus*, ending in *Cirrostratus*, with a lunar corona. 22. Morning cloudy and dark, by a large mass of smoke passing near us in the S.: rain, sleet: snow to the depth of several inches, with a very gentle breeze: moonlight evening. 23. Fair, much snow on the trees and shrubs: a strong westerly breeze: the rise of the barometer, like the previous fall, very sudden: at night stormy, with hail and much rain. 24. The snow mostly gone: elevated *Cirri*, with *Cumuli* in a pale blue sky: after *Cirrostratus* and haze at evening a gale through the night. 25. Much wind, with driving clouds: temperature 48° at nine, a.m. 26. a.m. Rain: snow in very large flakes: sleet: much water out since the late rains: rocky *Cumuli*, followed by *Nimbi* and gusts of wind, p. m.: clouds coloured at sunset. 27. Wet, a. m.: fine evening after a rainbow. 28. Elevated *Cirri* and *Cirrocumuli* stretching NW and SE: general obscurity followed, wivh showers and wind.

RESULTS.

Winds variable: in the latter part stormy from the Westward.

Barometer:	Greatest height	30·16 in.
	Least	28·85 in.
	Mean of the period	29·975 in.
Thermometer:	Greatest height	52°
	Least	20°
	Mean of the period	34·20°
Mean of the hygrometer		80·0°
Rain		3·10 in.

The evaporation for this and the preceding period taken together, is 1·38 inch.

METEOR.

A large and very luminous Meteor was observed *at two o'clock in the day*, the 6th of 2d Month, from Cambridge, and from Swaffham, Norfolk, descending from the zenith towards the northern horizon, within about twelve degrees of which it disappeared.—See a letter describing it, from Professor Clarke, in Thomson's Annals, &c. vol. ii. p. 273.

TABLE CXLI.

1818.		Wind.	Pressure. Max.	Pressure. Min.	Temp. Max.	Temp. Min.	Hygr. at 9 a.m.	Rain, &c.
3d Mo.	March 1	SW	29·64	29·62	47°	33°	59	18
	2	SW	29·71	29·52	48	30	65	17
	3	SW	29·35	29·25	50	32	82	14
	4	SW	28·83	28·50	47	36	72	22
	5	SW	29·20	28·83	44	34	60	—
	6	S	29·39	29·09	46	30	64	26
New M.	7	S	28·88	28·70	49	37	75	37
	8	W	29·33	28·88	47	31	58	3
	9	W	29·36	29·26	43	28	55	—
	10	W	29·42	29·24	42	29	60	—
	11	NW	28·84	28·80	44	30	54	—
	12	NW	29·26	28·84	45	32	62	36
	13	N	29·80	29·26	43	28	69	—
	14	SW	29·80	29·45	45	35	68	—
1st Q.	15	W	29·32	29·15	45	33	68	—
	16	NW	29·68	29·32	48	34	67	12
	17	W	29·98	29·68	52	37	58	
	18	SW	29·95	29·93	52	43	56	
	19	SW	29·95	29·58	52		59	—
	20	NW	29·88	29·58	45	28		
	21	SW	29·88	29·57	51	35	66	19
Full M.	22	SW	29·57	29·22	50	42	80	68
	23	SW	29·60	29·22	49	34	78	9
	24	SW	29·70	29·49	50	31	65	38
	25	NW	29·77	29·45	48	27	68	—
	26	Var.	29·96	29·15	39	33	77	55
	27	NE	30·26	29·96	45	27	64	1
	28	SE	30·26	30·12	49	35	63	
L Q.	29	S	30·13	30·05	49	27	53	
			30.26	28·50	52	27	65	3·75

NOTES.—Third Mo. 1. *Cirrocumulus* followed by *Cumulus* and *Nimbus:* squalls with hail and rain : the bow at nine, a. m. and again at three, p.m. 2. Fine, a. m. with clouds, as yesterday: wet, windy, p.m. 3. Hoar frost: fine, a.m.: clouds and wind, p.m. : rain, with large hail in the night. 4. Fine morning : the barometer, which had gone somewhat lower than here noted, rising abruptly : having risen about four-tenths, it took to falling again rapidly : there was a complete overcast sky (with haze in broad streaks, converging in the SW, and scud moving swiftly under it) till dark ; it now began to rain, and the wind rose to a greater degree of violence than for some years past, raging thus from SE and SW till past midnight ; when it abated, the barometer appears to have turned to rise more abruptly than before, having gone down an inch in 15 hours (the actual lowest point 28·35 inches) ; there is said to have been much thunder and lightning after midnight; the barometer fell, not uniformly, but by fits, at intervals of about a quarter of a minute, as the more violent gusts of wind came over. 5. Fine, a.m. : squally, with hail and rain, p.m. 6. The same. 7. Wet, stormy : much wind at night, with lightning far to the S and SE. 8. Fair, with *Cirrus* and *Cumulus :* the latter crossed with streaks of *Cirrostratus : Nimbi* succeeded, with showers: windy night. 9. Fair, with wind and *Cumulus :* the clouds assume a more tranquil aspect : a little snow this evening. 10. Some fine specimens of *Nimbus* to-day, from which a very little snow fell : clear night. 11. *Cirrus :* windy : in the night some rain, followed by snow from the N. 12. The ground covered with snow : *Cirrus,* followed by *Cumulus* and *Nimbus :* showers. 13. Rain, snow, and sleet, early : various modifications of cloud to-day : at night, a few drops. 14. Hoar frost : some rain, p.m. and evening : lunar halo. 15. Windy : some showers. 16. Various modifications of cloud, a.m. ending in *Nimbus,* and a shower with hail, p.m. : at night, calmer than of late, with a lunar corona. 17. Fair, a.m. : turbid sky above, with *Cirrus, Cirrostratus,* &c. : windy. 18. Close *Cumulostratus* most of the day : windy at night. 19. *Cumulus,* with *Cirrus :* windy, p.m. tending to S : a little rain in the night. 20. The clouds gradually thickened, as for rain ; but a brisk wind carried them off to the SE : hygrometer at five, p.m. 36°. 21. Hoar frost : the roads dusty : showers, p.m. 22. Some showers, with hail from NW : fine, p.m. : all night a hard gale from about SW with rain. 23. Morning wet, and stormy : fine, with clouds, (among which was the *Cumulus* capped), p.m. : night, pretty calm. 24. Fair, a.m. ; then a hail shower : much wind, with rain, in the night. 25. Some hail showers : large *Cumuli, Nimbi,* &c. 26. A steady rain from SE, with little wind, a.m. : the rain ceasing, p.m. the wind went by S to NW, and so probably by N to NE, where it was on the morning of the 27th, the barometer having risen rapidly with a uniform motion : a shower, p.m. 28. Fair, gloomy. 29. Fair, with *Cumulus* and *Cirrostratus..*

RESULTS.
Winds Westerly.

Barometer : Greatest height .... 30·26 in.
Least (observed)'....... . ... 28·50 in.
Mean of the period . ... 29·47 in.
Thermometer : Greatest height .. .. 52°
Least .... 27°
Mean of the period .... 39·70°
Mean of the hygrometer .... 65°
Evaporation ... . 1·10 in.
Rain . ..., 3·75 in.

Character of the period for the most part tempestuous, with frequent rains, the barometer running through a series of sharp depressions till near the close, when it suddenly assumed the elevation of fair weather. Almost all the showers, from the first, were more or less mingled with hail.

The storm of the 4th of Third Mo. appears to have been felt over the whole of South Britain, and to have been attended with much damage and loss of lives both by sea and land. I annex extracts from the papers, sufficient to shew its prevalence on many distant points of our island at the same time. In a letter which I received soon afterwards from Dr. William Henry of Manchester, its effects are thus mentioned : " On the evening of that day (the 4th,) we had a violent storm of wind from the SW, which did considerable damage. It began about eight, p. m. and continued with increased violence till three or four in the morning of the 5th. Near Macclesfield it was so violent, that the London mail to this town was detained there from eight, p. m. till five, a. m., and did not arrive here till ten hours after its usual time." The Barometer, it appears, fell on this occasion, at Manchester, to 28·20 in.; the highest *temp.* of the day 44°. The lowest point it attained with us (28·35 in.) was ascertained by means of the registering clock, from which I was not at that time in the practice of taking off the results for publication.

On the subject of the Barometer Dr. Henry adds, " My father once remarked it below 28 inches, and was greatly surprised that no rain followed. Several weeks afterwards he learned that a violent earthquake had been felt at that time, I think in Sicily." I introduce this remark, because it is probable most of my readers would be induced to attribute to a similar cause the phenomenon of a retrograde movement of the tide at *Hull*, which I annex, as I found it described in the public papers. It appears that in the afternoon of the 4th of Third month, the tide on that coast having turned to ebb, suddenly flowed again for a short time and to a greater height than before : in the early part of the morning of the 5th on the contrary, the depth of high water in the dock fell short considerably. Now, if we suppose this gale of wind arriving suddenly from the *Southward* (which by comparing the times of its beginning at Ryde, Isle of Wight, and at Manchester and Yarmouth, appears to be the fact,) a swell, produced by the compression of the water in the channel and straits of Dover, may have been propagated on the surface of the sea Northward, with greater swiftness than the storm could make its way (as far as regards its action on the face of the earth) across the land to Hull. The arrival of this swell at the critical time of the tide's turning accounts for the first fact. With regard to the second, it is matter of historical record, that an *off-shore* wind, (as this was at Hull,) if it blow long enough, and with sufficient force, may so remove the sea from the coast as to suspend a whole tide and give opportunity to the estuary of a considerable river to run itself empty: as has happened, I think more than once, to the Thames. Portsmouth and Hull were therefore placed on this occasion, by the operation of the same cause, in opposite circumstances: the one had " the highest spring tide ever remembered," and the other, a tide later, near five feet less water than was expected.

On Wednesday night (March 4th,) the Metropolis was visited by a storm more violent than any we remember for some years past. The wind began to blow from S to SW about eight o'clock, a perfect hurricane, accompanied with rain, and at times with lightning, and raged with increasing fury until near one, when a temporary abatement took place, which was followed by occasional squalls till between three and four o clock. We regret to learn, that this tremendous storm has been productive of considerable damage in various parts of the town.— *Pub. Ledger.*

Yarmouth, March 5.—A most tremendous gale of wind from the S to the SE, with rain, came on about eight o'clock last evening, which continued with increasing violence all night, and has done considerable damage to the shipping on this part of our coast.

Deal, March 5.—Last night it came on to blow a most tremendous gale from the South, and continued nearly the whole of the night with unabated violence; at midnight it blew a complete hurricane, accompanied with thunder and lightning, during of which, several vessels in the Downs suffered.

Portsmouth, March 5.—It blew a tremendous hurricane last night from S and SSE, accompanied with the highest spring tides ever remembered.

Ryde, March 5.—One of the severest gales of wind that has been felt here for the last thirty-seven years was experienced last night: it commenced about half past four p.m., and continued with increased violence until past eleven, during which time the greater part of the pier and several houses were demolished. The supposed damage is estimated at between four and five thousand pounds. No lives were lost, nor any damage done to the shipping.

Dartmouth, March 5.—We experienced a perfect hurricane last night at SSE, from six to ten p.m.

Exmouth, March 6.—On the 4th instant, between seven and eight o'clock in the evening, we had a most tremendous gale of wind about SSE, with dreadful rain, thunder, and lightning.

Falmouth, March 4.—At day-light this morning the wind was from the WSW, moderate, about eleven a.m. it backened to the S, and from that to SSE, and since that time to the present moment (ten p.m) it has blown a hurricane, with a heavy sea in from that quarter.

Penzance, March 6.—We had a very heavy gale here on Wednesday the 4th inst. from S to SSW·

Milford, March 6.—On the 4th inst. it blew a very heavy storm from SW to WNW.—*Shipping List.*

Leicester.—Wednesday night (Mar. 4) was one of the most boisterous recollected for years past; much damage has been sustained in this town, and many parts of the county.

Hull.—The following phenomena were observed here on Wednesday the 4th March. At high water, about 30 minutes past four o'clock in the afternoon, the wind then blowing from the SW, with moderate weather, the tide flowed at the Old Dock Gates 18 feet 6 inches. After the tide had fallen from one to two inches, the Dock Gates closed as usual with the ebbing of the tide, which then began again to flow to the height, as near as can be calculated, of four or five inches, thereby opening the gates again; and continued flowing a sufficient time to allow one ship and several small vessels to pass into the dock, before it again began to fall. The tempestuous night of Wednesday ensued; the wind blew a heavy gale still from the SW, and at high water, at five o'clock on Thursday morning, the tide flowed 14 feet 1 inch; being 4 feet 5 inches less than on the preceding evening; although from the spring tides having put in, the water ought, according to the usual state of things, to have flowed higher than on the Wednesday evening.

Plymouth, March 5.—At the commencement of the winter a few large stones were placed by themselves on the top or finished part of the Breakwater, to see if they would stand the winter gales: they stood all but this last, and this morning I found them washed from the top, and lying on the North Slope. There were three of them, one of *nine tons*, and the other two of five tons each: they will be replaced as soon as possible, for further trial.

Plymouth.—The effects of the late thunder storm of the 4th March, on a fir tree belonging to W. Langmead, Esq. at Elfordleigh, in the neighbourhood of Plymouth, are too singular to be omitted, and perhaps the most extraordinary ones that ever occurred in this county on such an occasion. The tree, in question, has been long admired for its size and noble proportions, being more than 100 feet high and nearly 14 feet in girth: but it exists no longer, having been literally shivered to pieces by the electric fluid. Some of the fragments lie 260 feet from the spot, and others bestrew the ground in every direction, presenting altogether a scene of desolated vegetation, easier to be conceived than described.—*Papers.*

1818.		Wind.	Pressure.		Temp.		Hygr. at 9a.m.	Rain, &c.
			Max.	Min.	Max.	Min.		
3d Mo.	March 30	SE	30·30	30·13	53	35	71	—
	31	NE	30·25	30·20	44	35	64	—
4th Mo.	April 1	NE	30·25	30·13	47	35	55	
	2	NE	30·32	30 23	47	38	64	
	3	NE	30·37	30·32	45	32	52	
	4	SE	30·32	30·02	49	22	50	
New M.	5	SW	30·02	29·22	54	40	63	4
	6	SW	29·67	29·22	51	34	73	—
	7	Var.	29·67	29·40	56	33	67	65
	8	S	29·40	29·24	63	48	61	25
	9	SW	29·50	29·24	56	40	61	6
	10	S	29·50	29·16	55	42	70	17
	11	NW	29·92	29·16	56	33	54	22
	12	W	30·02	30·00	47	27	55	2
1st Q.	13	SE	30·00	29 65	53	33	55	
	14	SE	29·69	29·63	59	28	53	
	15	SE	29·63	29·34	60	28	69	
	16	SE	29·34	29·20	57	47	52	16
	17	E	29·30	29 20	54	42	77	—
	18	NE	29·66	29·30	50	32	50	
	19	E	29·72	29·66	53	26		
Full M.	20	Var.	29·72	29 63	54	28	55	
	21	E	29·66	29·62	57	40	48	
	22	SE	29·62	29·45	58	43	53	—
	23	E	29·45	29·24	46	42	80	81
	24	NE	29·24	29·08	53	42	75	52
	25	SW	29·35	29·08	58	42	65	5
	26	Var.	29·35	29·25	68	52	60	14
L. Q.	27	SW	29·75	29·25	65	44	60	—
			30·37	29·08	68	22	61	3·09

NOTES.—Third Mo. 30. Hoar frost: a breeze, variable, succeeded by *Cirrus* mingled with *Cumulus:* a few drops, p. m. 31. Fine breeze: large *Cumulostrati*, with a few drops of rain: clear twilight.

Fourth Mo. 1. Fine: *Cumulus* passing to *Cumulostratus* in a brisk wind: at sun-set an evaporation of the clouds, followed by dew and an orange twilight. 2. Cloudy morning: *Cumulostratus* carried in a brisk wind through the day. 3. As yesterday, with the addition of red *Cirri* at sun-set. 4. Much sun, with long, faint, linear *Cirri*. 5. Hoar frost: calm: a warm sun, with much dust: *Cirrus* increased to obscurity in the evening, and it rained by night. 6. A gale through the day : calm night. 7. Wet, a. m. from the eastward : p. m. rain from the SW : a gale in the night. 8., Turbid sky : *Cirrocumulus* at nine, a. m. with the temp. 56° : afterwards, the wind southing, we had showers at intervals, and a gale by night. 9, 10, 11. Windy, with showers. 12. *Cumulostratus* chiefly, but with rain at intervals : in the evening the wind went to NW, with large *Nimbi*. 13. Fair, with *Cumulostratus*. 14. In the evening a large, faint, lunar halo, on a kind of *Cirrose* obscurity spread from NW towards the zenith. 15. Hoar frost : fine sky, with tendency to *Cirrocumulus*. 16. Hoar frost : *Cirrocumulus* by nine, a. m. in extensive beds : a smart breeze came on, with *Cirrostratus* and fleecy *Cumulus*, and *the first swallow* made its appearance about five, p. m. : rain ensued after dark, with a fragrant smell from the turf. 17. Drizzling morn : fine day. 18. Windy, overcast, bleak morning : fine day. 19. *Cumulostratus*, windy. 20. Hoar frost : fair, with clouds ; in the evening a westerly current was evident above, by the motion of elevated *Cirrostrati :* two different beds of this cloud had appeared at sun-set, crossing at an oblique angle in the S : the eclipse of the moon was well seen at intervals through these. 21. Little wind : fair. 22. Wet, p. m. 23. Very wet, a. m. and again, with wind, at night. 24. Overcast day : wet evening and night. 25. Some drizzling rain, after inosculation of heavy *Cumuli* with a stratum of clouds above. 26. Fleecy *Cumuli*, with *Cirri*, and tendency to *Nimbus* in the S : at nine, a. m. an unusual agitation, evidently electrical, was produced in a *Cirrus*, by the passage beneath it of fleecy *Cumuli*, which came from S, with the vane at E : thunder clouds soon after formed, and before one p. m. we heard three distinct explosions ; two successive showers of rain mixed with hail followed, but without wind : in the evening, large thunder clouds continuing about, it lightened for some hours in the distance, nearly all the horizon round, the W only being free from it : the wind SE. 27. Cloudy, wet morning : windy at SW : fine afterwards, with large *Cirrus* above *Cumulus :* some lightning at night in the NE.

RESULTS.

Winds Variable, with much South East.

Barometer: Greatest height .... 30·37 in.
    Least .... 29.08 in.
    Mean of the period........... 29·642 in.

Thermometer: Greatest height .... 68°
     Least .... 22°
     Mean of the period........... 45·36°

Mean of the hygrometer............... 61°

    Evaporation. .... 1·40 in.
    Rain........ .... 3·09 in.

The excessive rains continuing, have occasioned repeated overflowings of the river Lea into the marshes. Vegetation, which continued nearly dormant at the commencement of this period, was making considerable progress towards the close of it.

———

A *Tornado* occurred in the county of Middlesex at the time of the thunder-storm, which I have noticed (Fourth Mo. 26,) as preceded by an unusual demonstration of electricity in the clouds. It was observed by Col. Beaufoy from Bushey Heath, near which its course terminated, after proceeding in a direct line about five miles in 20 minutes, but probably moving in its circular whirl with more than five times that velocity. In its course it uprooted trees, unroofed houses, threw down walls, and in short removed every thing that impeded its progress; being visible by means of the inverted cone of cloud occupying its interior. See Thomson's Annals of Philosophy, vol. xi. p. 442, &c.

———

A tremendous storm of hail and rain, accompanied by thunder and lightning, fell at Hampstead on Sunday, April 26, at half-past eleven, and the effects were severely felt at Hendon and the adjacent villages. Upwards of twelve large trees were blown down at Hendon. At one o'clock its effects were felt at Dunstable, Redburn, and Market-street, where the hailstones were so large as to break the church windows, &c. The accounts from various parts mention the dreadful effects of this storm.—*Pub. Ledger.*

A Letter from *Boston*, dated April 28, says,—From eight until eleven o'clock on Sunday night, this town and the neighbourhood were visited by a most dreadful storm of thunder and lightning; when the latter was more extensively diffused than any that has been witnessed in this part of the kingdom since the month of July 1809. An old man, who has travelled in the East, says, ' He has seen such lightning before in India frequently—in England never.'—*Pub. Ledger.*

Hail Storms in France.

The French Journals continue to give deplorable accounts of the ravages in the departments of the Saone and Loire, occasioned by dreadful hail storms. The hail-stones are described as being as large as pullet's eggs. The game was every where found dead in the fields; and several persons were severely wounded by the hail. In one vineyard alone, between Orleans and Beaugency, the damage done by the storms is estimated at 4000 pipes of wine.—*Pub. Ledger, May* 16.

Ice in the Atlantic.

Capt. Quereau, of the Grand Turk, which has arrived at Derry from New York, states, that on the 15th of February, in latitude 43, he passed through several islands of ice, some miles in extent, and from 3 to 400 feet high The ship, with strong westerly gales, was two days in getting clear of them.—*Pub. Ledger.*

The Speedy Packet, arrived from New York, saw, on 28th March, in lat. 45. 32. long. 46. 2 an island of ice, the summit of the peak of which was, on a moderate calculation, 400 feet above the surface of the water; and in lat. 45. 36. long. 48. 00. W. passed two more, the largest of which, was 250 feet above the sea and three quarters of a mile in length.— *Shipping List.*

The ships arriving from the westward continue to report having seen immense masses of ice. The Minerva, from New York. in lat. 41. 50. N. lon. 50. W. passed a number of islands of a height of from 150 to 200 feet. The Washington too saw several fields of ice, in lat. 42. W. long. 48. 20.—*Pub. Ledger, May* 25.

	1818.	Wind.	Pressure. Max.	Min.	Temp. Max	Min.	Hygr. at9a.m.	Rain; &c.
4th Mo.	April 28	SW	29·98	29·80	53	30	53	
	29	NE	29·98	29·75	65	41	50	
	30	Var.	29·88	29·75	53	48	70	51
5th Mo.	May 1	SW	29·88	29·70	60	35	56	5
	2	Var.	29·85	29·45	69	47	52	39
	3	SW	29·48	29·36	66	43	75	3
	4	NW	29·48	29·40	67	42	58	
New M.	5	NE	29·40	29·26	65	48	62	18
	6	NE	29·26	29·22	65	45	72	31
	7	SW	29·51	29·26	62	43	63	—
	8	S	29·51	29·40	65	44		1·46
	9	SW	29·84	29·51	60	45		2
	10	SW	29·84	29·67	64	37	50	—
	11	S	29·67	29·57	67	49	43	2
	12	NW	29·63	29·45	66	40	44	15
1st Q.	13	SE	29·45	29·30	61	41	59	10
	14	S	29·47	29·30	58	39	57	1
	15	NW	29·52	29·47	63	41	46	—
	16	N	29·65	29·52	63	50	45	5
	17	NW	29·87	29·65	61	44	75	
	18	N	30·00	29·87	69	47		
	19	NE	30·05	30·00	57	38		
Full M.	20	E	30·20	30·05	65	45		
	21	E	30·23	30·20	59	36		
	22	E	30·32	30·23	61	42		
	23	NE	30·35	30·32	61	37		
	24	SE	30·35	30·25	63	44		
	25	NE	30·26	30·25	65	40		
L. Q.	26	NE	30·33	30·23	69	42		
			30·35	29·22	69	30	57	3.28

NOTES.—Fourth Mo. 28. Much dew: at nine a. m. a brisk wind carrying *Cumuli*, above which appeared beds of *Cirrus* and *Cirrocumulus*, moving from SE: a fine day ensued, with *Cumulostratus*. 29. Fine. 30. Overcast early, with the wind NE; after which wet till evening.

Fifth Mo. 1. A fine day, save a shower or two. 2. Large *Cumuli* rose, which, in the E especially, mingled and inosculated with *Cirrostratus* above; I suspected thunder in that direction: at sun-set, *Cirri* from N to S, above *Cirrostrati* ranging E and W: rain by night. 3. Drizzling, a. m.: fine, with *Cumulus*, and *Cirrus* at mid-day: in the evening, heavy showers appeared to the N and NE, with much

Cirrostratus overhead. 4. Very fine, with *Cumuli*, and large, plumose *Cirri* stretching E and W; the clouds, though heavy, dispersed at sun-set. 5. Sunshine at six a. m., with a few *Cirri*, &c.: before seven, a sudden mist came on from the E and NE, which obscured the view of the *Solar Eclipse* during the middle half hour of the time; the dew lay on the grass till noon, in the sunshine, and large *Cumuli* formed, inosculating with the clouds above: at two p. m. some heavy showers fell, but so local, that the road, half a mile off to the S, remaimed dusty: in the evening, *Nimbi* appeared in thunder-groups to the SE and S, and, finally, more extensive rain came on, with the wind SE. 6. Rain, a. m., and at night.

RESULTS.

Winds variable.

Barometer:	Greatest height	30·35 in.
	Least	29·22 in.
	Mean of the period	29·766 in.
Thermometer:	Greatest height	69°
	Least	30°
	Mean of the period	52·84°
	Hygrometer (mean of 18 days)	57°
	Evaporation	1·70 in.
	Rain	3·28 in.

Having left home on a journey on the morning of the 8th of Fifth mo. I did not witness a very uncommon fall of rain which took place in this neighbourhood. It commenced early in the evening of that day, and lasted about 12 hours. Near an inch and a half of water descended in the above space of time, which, taking the shortest course from the higher ground to the hollows, filled the latter several feet deep, and overflowed the roads, in several places not usually subject to this accident. Much inconvenience, and some loss of property, ensued, the particulars of which were detailed in the newspapers of the subsequent days. This heavy rain seems to have been connected with a change in the general current, which, after a few days further continuance of unsettled weather, became established from the northward, the barometer assuming a high level, and the earth drying rapidly. It was, indeed, a singular spectacle to behold the ground saturated with water, and every spring running, up to so late a period in the season as the middle of the fifth month, when our fields are commonly dry enough, in every situation, to admit of the soil being pulverised by the harrows.

Weather Abroad.

About the end of April parties in sledges were still making at *Stockholm*; at *Petersburg* the people were walking and driving carriages on the ice of the Neva; while but a few days later they were complaining of the heat at *Vienna*.

According to a letter of the 10th ult. from *Lisbon*, the weather there had been for some weeks excessively cold and rainy.—*Pub. Ledger*, June 1.

1818.		Wind.	Pressure. Max.	Min.	Temp. Max.	Min.	Hygr. at 9a.m.	Rain, &c.
5th Mo.	May 27	NE	30·23	30·13	67	39		
	28	NE	30·13	30 09	69	44		
	29	NE	30 13	30·01	63	41		
	30	NE	30 02	29·90	65	33		
	31	NW	29 97	29 90	74	51		
6th Mo.	June 1	NW	30 00	29 97	77	57		
	2	W	30·05	30 00	80	43		
New M.	3	W	30·14	30 05	80	45		
	4	SE	30 30	30 14	82	45		
	5	E	30·33	30·30	79	43		
	6	NE	30·33	30·25	78	45		
	7	SE	30·27	30 18	77	52		
	8	SE	30 27	30·23	75	49		
	9	SE	30·23	30 20	75	46		
	10	SE	30·21	30 10	80	50		
1st Q.	11	SE	30·10	29·98	84	47		
	12	E	29·98	29 85	88	51		
	13	NW	29·93	29·82	89	58		3
	14	NW	30·07	29·93	75	49		—
	15	SW	30·05	29 92	78	55		—
	16	SW	29 92	29·75	78	59		
	17	SW	29 70	29 67	74	54		6
Full M.	18	NW	29·79	29·69	75	49	50	—
	19	SW	29·79	29 53	72	52	45	2
	20	SW	30 00	29·60	72	46	47	1
	21	SW	30·0u	29·75	71	56	43	6
	22	SW	29 76	29·64	71	50	62	25
	23	NW	29·87	29 76	71	53	44	
	24	NW	30 09	29·77	74	52	47	—
L. Q.	25	SW	30 09	30 02	79	56	52	
			30·33	29·53	89	33	48	0·43

NOTES.—Sixth Mo. 6. Since this period came in, the weather has afforded little variety. The days have been serene, with breezes, which commonly increased with the temperature, and died away at sun-set: the nights nearly calm, with dew, and a peculiarly clear, but not high-coloured twilight. Thunder clouds have shown themselves at intervals in the horizon; and to-day there are large plumose *Cirri*. 8. My brother observed, about nine, p. m. a bright, blue meteor descending from the zenith to the NW. 10. After sun-set, some beautiful diverging shadows on a pure, dilute, carmine tint in the NW. 11. Thunder clouds about. 12. A thunder group in the N and NW: the *Cirrostratus* for a short time assumed the form of the *Cyma*, and several discharges were heard while the *Nimbi* expanded their crowns within view: after this, it lightened in some clouds to the SE. 13. *Cumuli*, mingled with haze and *Cirri*, were followed in London by a smart thunder shower; while at Tottenham there fell but little rain: a lunar corona ensued. 14. A little rain, a. m.: a large, faint lunar halo. 15. A few drops at evening. 16. Cloudy: a strong breeze. 17. A light gale, with a rainy sound, and much cloud; but the showers proved scanty. 18. Much cloud, chiefly *Cumulostratus*: after some light showers, and appearances of rain and thunder to the southward, the twilight cleared up orange. 19, 20. Windy, cloudy: light showers; *Cumulus*, *Cirrocumulus*. 21. *Cumulus*, with the lighter modifications above, increased to obscurity: wind through the day, and small rain, evening. 22. Windy, cloudy morning: this day more decidedly showery. At eleven p. m. a shooting star descended to the SE. 23, 24. More calm, with summer clouds in various modifications. 25. A very slight rain, a. m. followed by fine blue sky, and various clouds carried by a strong breeze.

RESULTS.

Winds in the fore part light and Easterly, in the latter part Westerly and stronger.

Barometer: Greatest height .... 30·33 in.
Least .... 29·53 in.
Mean of the period .... 29·998 in.

Thermometer: Greatest height. .... 89°
Least .... 33°
Mean of the period (at the Laboratory) 62·36°

Mean of the hygrometer (the latter week) 48°

Evaporation (a few days estimated) ... 4·50 in.
Rain .... 0·43 in.

The clear hot sunshine of the greatest part of this period had the effect of establishing the summer in our climate in a manner to which we have long been unaccustomed. The deeper green of the foliage and the richer colour of many flowers in particular presented a striking contrast to their appearance during the last two seasons; while the soil, parched and cracked over the whole surface of our loamy meadows, bore ample testimony to the continued receptive power of the atmosphere. Yet (to use a familiar phrase) *the turf did not burn*, probably in consequence of the supply of moisture still left at a certain depth in the soil.

Ice Islands in the Atlantic.

The Albion, Davis, arrived at Amboy in 45 days from North Wales, saw, June 11th to 18th, from lat. 46. lon. 46. to lat. 43. lon. 52. upwards of 40 islands of ice, many of them very large.—*Pub. Ledger.*

1818.		Wind.	Pressure. Max.	Min.	Temp. Max.	Min.	Hygr. at 9 a.m.	Rain, &c.
6th Mo.	June 26	SW	30·05	29·95	79	49	47	
	27	SE	29·95	29·67	84	55	43	—
	28	SW	30·17	29·75	72	52	44	20
	29	Var.	30 26	30·17	81	51	43	—
	30	NW	30·26	30·15	84	52	43	—
7th Mo.	July 1	NE	30·15	30 02	81	52	42	—
	2	Var.	30·22	30·10	73	44		
New M.	3	NW	30·22	30 06	79	57	45	
	4	N	30·06	30·04	77	52		
	5	NE	30·10	30·04	79	51	46	
	6	SE	30·10	30 00	84	52	45	
	7	SE	30·00	29 80	81	56		
	8	NW	30·11	29·80	74	50	46	
	9	NW	30·11	30·10	78	53	45	
	10	Var.	30·10	29·95	76	55	42	
1st Qr.	11	SW	29·95	29·76	79	50	42	33
	12	NW	29·85	29·76	74	57	70	1
	13	N	30·20	29·85	77	52	52	
	14	NW	30·32	30·20	83	57	52	
	15	Var.	30·32	30·28	86	53	47	
	16	NE	30·28	30·18	88	62	45	
Full M.	17	E	30·20	30·09	82	52	53	—
	18	SE	30·09	29·91	84	57	50	
	19	NW	29 95	29·88	85	59	50	
	20	Var.	29·95	29·92	76	52	55	
	21	NW	30·05	29·95	80	56	45	
	22	SW	30·13	30·05	84	55	45	
	23	SE	30·13	29·83	83	60	47	
	24	SE	29·83	29·80	93	61	40	9
			30·32	29·67	93	44	47	0·63

Notes.—Sixth Mo. 27. It is said to have been misty early. Some remarkable, rapid changes in the electrical state of the clouds took place, the wind being brisk, veering from SE to SW. *Cirri*, passing to *Cirrocumulus* and *Cirrostratus*, grouped like the ribs of a vessel, on a kind of keel presenting downwards; very dense and magnificent. With these were mingled the rudiments of *Nimbi*, one or two of which formed in sight, and probably discharged to the NE of us : a few drops fell, and there were distant thunder storms in different directions at night. 28. Some fine rain, a. m. : several short, heavy showers about noon : inosculation, and gray sky, evening. 29, 30. Fine, with large *Cirri* above *Cumuli*: some drops of rain.

Seventh Mo. 1. A fine display of *Cirrocumulus*, with a specimen of *Cirrostratus* resembling the grain of wood: also large plumose *Cirri*, p.m. : *Cumulostratus*, and a few drops, evening. 2, 3. Exhibitions of the lighter modifications variously interchanging and mingling, succeeded by *Cumulostratus*. 4. Windy morning, overcast with *Cumulostratus:* a fine day: twilight coloured, with diverging shadows. 5. Very fine day: *Cirrocumulus* above *Cumulus* producing beautiful clouds by inosculation. 6. At *three* this morning, in the NE, a most extensive orange twilight, in the form of a pyramid, resting on a base of low purple haze, occasioned by dew in that quarter. A fine day ensued, with a breeze, and *Cumuli* casting shadows in a somewhat hazy air. 7. The shadows radiating downward from clouds continue, perhaps occasioned by fine dust floating. I observed, in passing Hounslow Heath, two whirlwinds, carrying the dust in a narrow perpendicular vortex to a great height in the air, from whence it perceptibly showered down again. 8. *Cumulostratus*, after a clear morning : strong breeze and much cloud, with a few drops. 9. Clear morning, with *Cumulus, Cirrus*, and a breeze. About seven, p. m., setting out to return from London, I saw, in the NW, a remarkably large *Cirrus*, composed mostly of straight, diverging fibres, extended towards the SW ; and which, when I got home, had passed to *Cirrostratus*. In this cloud (as it appears) my family at the same time observed a coloured *solar halo* with two rather indistinct *parhelia*, the whole of which had escaped my notice in coming out of town. 10. A few large drops between six and seven, a. m. : close *Cumulostratus* prevailed afterwards. 11. Large *Cirri*, passing to the form of the *Nimbus*, mingled with *Cirrocumulus* and *Cirrostratus*. In the evening an extensive obscurity in the W and SW, fronted by dense *Cirrostrati*: a fresh, turfy smell came with the wind, and at length, at half-past ten, it began to rain steadily with us. 12. Wet morning : fine day afterwards. 13, 14. Fine, with *Cumuli*, &c. dew, and orange twilight. 15. A *Stratus* last night : thunder-clouds about : the moon bright gold colour, crossed by fine streaks of *Cirrostratus*. 16. The moon paler, amidst hazy *Cirrus* and *Cirrostratus*, &c. in SE. 17. Cloudy morning : light shower, then fine with *Cirrus* and *Cirrocumulus*. 18. Thunder clouds, p.m. : *Nimbi*, &c. grouped in the N. 19. Wind SE ; thunder came within hearing to the NW, p. m. : temp. 85° : hygrometer 30° : not a drop of rain here, and wind NW after it. 20. Thunder groups, and rain visible to the northward : fair with us : clouds red at sun-set. 21. Wind W, a. m. *Cirrocumulus*, chiefly in

strips from N to S; then *Cumulostratus,* &c.   A very variously com-
pounded and coloured sky during twilight.   22. Fleecy *Cumuli,* &c.
a.m. :   23. Serene, with *Cirrus,* and fine breeze.   24. *Cirrus* and
Cirrocumulus proceeding to electrical formations : strong breeze and
slight solar halo : p. m. after the maximum of temp. was over, *Nimbi,*
with thunder and lightning, approached from the south.   The clouds
at sun-set showed very rich crimson, lake, and orange tints ; and we
had showers, with a hollow wind, and lightning, till past midnight.

RESULTS.

Winds light and Variable.

Barometer : Greatest height. . . .   . 30·32 in.
Least . . . . 29·67 in.
Mean of the period . . . . 30·037 in.
Thermometer : Greatest height . . . . 93°
Least . . . . 44°
Mean of the period (at the Laboratory) . . 67·24°
Mean of the hygrometer . . . . 47°
Evaporation . . . . 4·60 in.
Rain. . . . . 0·63 in.

A period unequalled in warmth since the year 1808, when the
seventh month averaged 67·19°, and the thermometer at Plaistow rose
to 96°.   The eighth month, 1802 (averaging 67·56°), is the only one
that has *exceeded* the present in heat for 20 years past.

THUNDER STORMS.

Devonshire, July 19.—The villages of Lympstone, Exmouth, Woodbury, Otterton,
Budleigh, and other places in that neighbourhood, were visited with a sudden and
most tremendous storm of rain, accompanied with thunder and lightning.

July 19 a dreadful thunder-storm burst over *Stafford.* The lightning set fire to
a barn at the farm of Mr. Lathbury, adjoining that town. The barn was burnt
to the ground, before the progress of the flames could be arrested ; and it was
with difficulty that the rick-yard and house were preserved. The lightning struck
the premises of Mr. Nixon, hatter, one of the Society of Friends, at the same time.
The same evening a tremendous storm of rain burst over the village of *Brereton,*
and the roads were momentarily impassable from the sudden accumulations of
water.

The same day there was a heavy tempest at *Downham Market*, Norfolk, and in the neighbourhood; and, while the rain poured down in torrents, a ball of electric fire fell and exploded upon a new barn at Magdalen, belonging to Mr. E. Butrick, which consumed the same, together with upwards of 20 sacks of rye grass-seed, in the sacks. We have not heard of any other damage.

A most beautiful water-spout was seen at *Gainsborough*, in Lincolnshire, on Monday last, so awfully grand, as to astonish every beholder of its majestic movements and colours.—*Pub. Ledger.*

For three days lately *Gloucester* has been visited by very heavy storms of rain, accompanied by thunder and lightning; on Saturday evening, particularly, the lightning was exceedingly vivid, and the crashes of thunder awfully loud and sublime. The mast of a barge lying in the basin of the canal, was struck by the electric fluid, and splintered from the top to the bottom, a solid piece being carried away from the lower part. There were seven persons on board at the time; but though most of them felt the shock, and a handkerchief round one of their hands was burnt through by the lightning, yet none of them suffered injury.—*Pub. Ledger*, *July* 22.

HEAT ABROAD.

HAARLEM, *July* 31.—A good deal of damage has been done within the last week by terrible storms of thunder and lightning. Five or six farm houses in different parts of the country were struck by the lightning and totally consumed.

Accounts from *Arnheim* say—that on the 24th and 25th the heat was there from 92 to 94 of Farenheit. An equal degree of heat prevailed in those days in many parts both of Holland and Brabant —*Pub. Ledger.*

1818.		Wind.	Pressure. Max.	Min.	Temp. Max.	Min.	Hygr. at 9 a.m.	Rain, &c.
7th Mo.	Full M. July 25	SW	29·85	29·80	77°	54°		
	26	S	29·85	29·80	84	62	45	2
	27	N	30·22	29·80	79	51	52	
	28	SW	30·27	30·22	72	47	37	
	29	SW	30·27	30·16	81	56	40	
	30	SW	30·16	30·10	82	59	40	
	31	W	30·10	29·97	80	58	46	12
8th Mo.	Aug. 1		30·18	30·03	70	50	52	
	L. Q. 2	S	30·18	30·10	70	43	50	
	3	S	30·10	30·07	79	47	48	
	4	SE	30·10	30·05	87	50		
	5	E	30·03	30·00	93	57		
	6	N	30·09	30·03	88	59	40	—
	7	NW	30·10	30·07	76	52		
	8	NW	30·10	29·95	78	53		
	New M. 9		29·95	29·87	82	56	52	
	10	NE	30·20	29·95	72	43		
	11	NE	30·20	30·1C	70	50	47	
	12	NE	30·13	30·07	72	46	50	
	13	NE	30·13	30·10	76	45	47	
	14	NE	30·10	30·08	71	53		
	15	NE	30·11	30·08	68	53		
	1st Q. 16	N	30·08	30·00	76	46		
	17	N	30·00	29·90	76	45		—
	18	NW	29·94	29·88	77	50		
	19	N	30·04	29·94	66	46		—
	20	N	30·04	30·03	66	50		
	21	NW	30·06	30·00	71	44		
	22	NW	30·20	30·06	66	43	39	
			30.27	29·80	93	43	45	0·14

NOTES.—Eighth Mo. 4. With the exception of a gentle rain in the evening of Seventh month 31, steady fine weather has continued. Much *Cirrocumulus* has appeared of late. This day, in travelling, I observed the clouds, both at sun-rise and sun-set, beautifully coloured with a double gradation of tints, in which the successive effects of the direct and refracted rays were very distinctly marked. 6. Wind in the morning, SE, brisk, with *Cirrostratus* and *Cirrocumulus*; the latter formed in one instance out of *Cirrus* with unusual rapidity: the wind veered gradually from SE by SW to NE: at nine, p.m. a strong breeze blowing, with an appearance of rain to NW, it began to lighten : at first, a very faint blue flash ; then others, gradually increasing in intensity at intervals of about a minute, filling the air, without being referrible to any point of the compass, followed generally by a sudden puff of wind, and without thunder. In twenty minutes, however, thunder began to be heard in the W and NW, and a storm passed in view to the NE, the flashes broad and vivid on the whole North horizon, and crossed by delicate striæ of a different colour. We had only a few drops of rain, and it was over in two hours. 7. The sun-set was more richly coloured with yellow (passing at length through orange to lake and purple) than I remember ever to have seen it in this tint before. It literally glowed like a bright flame on the lower surface of some dense *Cirri*, passing to *Cirrocumulus*; which modi-fication was well marked afterwards. 9. A fine coloured sun-set again, but in deep orange passing to red, and succeeded by *Cirrostratus*. 10. Cloudy, with a brisk wind most of the day : *Cirrostratus* and dew. 11 — 13. Fine breeze, varying to N and E: much dew; twilight orange ; and the moon pale. 14—22. Pretty strong breezes prevailed during this interval : the sky presented usually the *Cumulus* passing to *Cumulostratus ;* but at intervals this modification took its character from *Cirrocumulus*, which entered into its composition from above. There was scarcely any *Cirrus*, or obscurity above the clouds, but rather a cold, transparent blue : two or three times the density of the clouds promised showers, but it always ended in a very light sprinkling. Coloured skies at sun-set were frequent; as also the appearance of diverging bars of light and shade, which I ascertained in several instances to be due to the immense quantity of *dust* constantly floating in the air. 22. This morning, being gray with *Cirrocumulus* and very cool, seemed like the commencement of autumn ; and the warmth of a fire was acceptable in a north room in the evening.

RESULTS.

Winds Southerly in the fore part; Northerly, with depression of temperature, in the latter part of the period.

Barometer : Greatest height .... 30·27 in.
           Least .... 29·80 in.
           Mean of the period .... 30·051 in.
Thermometer : Greatest height....... .... 93°
           Least .... 43°
           Mean of the period .... 63·32°
    Mean of the hygrometer .... 45°
           Evaporation, nearly .... 4 in.
           Rain .... 0·14 in.

A period unequalled in *dryness* since the beginning of 1810 ; when, with a *frosty* air, under a similar course of winds, and the barometer averaging 30·07 in. there fell in 30 days only 0·12 in. of rain.

————————

Whirlwinds, &c.

(Copy of a Letter to the Author.)

Sir,

By your request I once had the pleasure of sending you an account of a luminous meteor seen by me at Peckham. Founded on this, and your curiosity in all the phenomena of our atmosphere, I presume the following may be acceptable. While sitting at a window facing the south, I observed a column of dust rising from a field that had been roughly ploughed the day before. I was surprised at the size of the column, as also at the figure that was preserved—that of a truncated cone, the base upwards, its diameter about 13 feet, its height perhaps 30 or 40 feet, and the diameter of the part next the ground rather less than a yard. At first only dust rose within it, but after about half a minute, large pieces of earth were lifted, I suppose about ten or twelve inches from the ground, one of these could not be less than the size of a man's head; this rose within the centre of the whirl. There was scarcely any wind at the time, but it moved generally in the direction of the current (at that time from the SSW,) with perhaps about half its velocity, The *largest* pieces raised (as well as the smallest) moved spirally, but these were not carried many inches. It travelled about 300 yards, and I then lost sight of it behind some trees: it was at that time increasing in height. I was within two hundred yards of it when it raised the heaviest pieces. It happened I should have told you at about noon on Tuesday, August 4, the day very hot, Fahrenheit's Thermometer at about 89° in the shade. Numerous small Cirro-cumulus and dense Cirrus clouds at a very great elevation, thin streamers directed to NW, N, and NE.

The above, if at all interesting, is much at your service.

I remain, respectfully, yours,

Peckham, Aug. 5, 1818.   John Wallis.

A Tornado passed over the village of *Howell*, in Lincolnshire, last week. It appeared like a body of smoke, was preceded by a small black cloud, passed very near the earth, and completely unroofed a low building, and tore the boughs from the trees as it passed, and carried them a considerable way. Coming in contact with a large ash tree, it split a piece from the trunk 12 feet long, and as thick as a man's body, carrying it at least 100 yards.—*Pub. Ledger, Aug.* 8.

On Friday evening (Aug. 7,) when a moment before there appeared to be a perfect calm, the inhabitants of *Croydon*, near the church, were thrown into alarm, by the sudden rising to a great height in the air, of fourteen pieces of cloth that had been pegged down in the bleach ground of Messrs. S. and T. Starey, one piece of which was so twisted round the steeple of the church, that it required a great length of time to disentangle it.

Scotland.—A singular and very beautiful phenomenon took place in the atmosphere on Saturday se'nnight soon after the commencement of the thunder storm, immediately above the Clyde, and a little to the west of Mauldsie Castle. At one o'clock the clouds in the atmosphere seemed to rush with much rapidity to one common centre, and soon after, a large inverted cone was formed, hanging perpendicular to the horizon, which moved round upon its centre with great velocity, for upwards of six minutes. It then changed its position, and from perpendicular became horizontal; and what had formed the lower point of the cone, formed a figure like a large wheel, and rapidly turned round upon its axis, throwing off, as it were, large flakes of transparent white clouds, like wool or cotton. After moving in that position for six or eight minutes, the motion decreased, and the cone was absorbed in the contiguous clouds. A very black and dense cloud had continued moving in a south direction, to the place of attraction, during the whole time of this phenomenon (which was in all about half an hour) and very soon seemed to lose the cohesion necessary for keeping it in a body, and was seen falling down in large torrents of rain. Immediately before this phenomenon was observed, a large fire ball was seen to dart from the atmosphere, near the part which became the place of attraction.—*Edinburgh Paper.*— *Pub. Ledger, Aug.* 12.

1818.		Wind.	Pressure. Max.	Pressure. Min.	Temp. Max.	Temp. Min.	Hygr. at9a.m.	Rain, &c.
8th Mo.	L. Qr. Aug. 23	N	30·20	30·15	66	58	50	
	24	SW	30·15	30·07	71	56	42	
	25	NW	30·07	29·95	72	50	45	
	26	NW	29·95	2·75	75	55	47	
	27	W	29·75	29·60	68	60	43	10
	28	NW	29·94	29·65	73	53	52	
	29	W	29·94	29·86	80	58	48	
New M.	30	W	30·00	29·86	76	40	52	—
	31	E	30 00	29·49	75	55	50	
9th Mo.	Sept. 1	SW	29 70	29·49	74	49	50	2
	2	SW	30·07	29·70	71	50	52	—
	3	SW	30·05	29·92	71	61	48	—
	4	S	29·99	29·96	75	63		9
	5	W	29·96	29·70	68	55	66	95
	6	W	29·85	29·70	69	55	65	2
1st Q.	7	NW	29·92	29·85	64	45	46	
	8	Var.	29·90	29·60	65	40	60	—
	9	NW	29·68	29·59	63	43	65	1
	10	NE	29·85	29·68	61	39		
	11	NW	30·00	29·85	60	42		
	12	NW	30·20	30·00	66	48		
	13	N	30·30	30·20	68	41		
Full M.	14	SW	30·25	29·75	67	58		
	15	SW	29·75	29·60	59	43	50	25
	16	NW	30·10	29·60	56	39	60	12
	17	N	30·20	30·05	57	41	57	
	18	NW	30·05	29·85	63	51	72	—
	19	S	29·85	29·58	67	50		2
	20	SE	29·58	29·38	61	44	60	—
	21	SE	29·63	29·32	69	49	70	13
			30·30	29·32	80	39	54	1·71

NOTES.—Eighth Mo. 23. Morning very clear: mid-day, *Cumulus* beneath large *Cirri:* p. m. inosculation, followed by a shower to the NW, which sent us a turfy odour with the wind. 24. *Cirrostratus*, followed by *Cumulostratus*, at times heavy; the wind veered to SW, p. m. 25. Large *Cirri*, directed from SW to NE. 26. *Cumulostratus* and *Cumulus* crossed by *Cirrostratus*. 27. The hygrometer advanced to 67°: gentle rain, a. m.: cloudy, p. m. 28. *Cumulus* and *Cumulostratus*: a little rain, evening. 29. *Cirrocumulus*, beautifully coloured at sun-set, in lake shaded with violet. 30. Some very light rain, a. m.: fair, with fresh breeze after it. 31. Large plumose *Cirri*.
Ninth Mo. 1. Lowest temperature on the ground 44°. This morn-

ing from two to three, it thundered and lightened much to the SE: thunder clouds prevailed a. m.: wind SE, and a little rain: a slight shower again at night, and much dew after it: the hygrometer advanced to 80°. 2. After large *Cirri, Cumulostratus,* which inosculated about sun-set with a scanty *Cirrocumulus.* 3. A mixed sky, with a slight driving shower at evening: cloudy night. 4. A sweeping rain, early: hygrometer 80° at six, a. m.: much hollow southerly wind: *Cirrocumulus,* followed by ill-defined *Cirrus* with *Cumulus;* and about five, p. m. a *Nimbus,* shaped like a low circular hay-rick, with a capped *Cumulus* by its side, on the NE horizon. 5. Much rain, for the most part small and thick. 6. Wet, cloudy morning: very turbid sky: hygrometer at 80°: calm air: in the evening, inosculation of *Cumulus* with *Cirrocumulus;* after which frequent lightning between nine and ten, p. m. 7. Morning gray, with *Cirrocumulus:* sun-shine followed, with inosculation of *Cumulus* and *Cirrostratus.* 8. Large *Cirri,* with fleecy *Cumuli:* the latter attached themselves in their passage to the smoke of the city, and appeared to disperse downwards into it. Thunder clouds followed this appearance, and a smart storm passed in the S, from W to E, about five, p. m.: the crown of the nearest *Nimbus* reached our zenith, and we had a few drops; while it rained hard, with a bow in the cloud, within two miles of us. 9. Heavy *Cumulostratus:* and showers, p. m. 10. Fine breeze, with *Cumulus* and *Cirrus;* the latter survived the sun-set, and was kindled by the refracted rays with flame colour passing to red: calm at night, with hygrometer 45°. 11, 12, 13. Chiefly *Cumulus,* and *Cumulostratus* by inosculation: some fine grouping of the clouds at intervals: large *Cirri* at the conclusion. 14. A large meteor seen passing northward: windy night. 15. Cloudy, windy: hygrometer, 75°: wet, p. m. 16. Much dew: a rapid propagation of *Cirrus* from the S, followed by *Cumulostratus* and showers: during a heavy shower about nine, p. m. it thundered in the NW: the barometer stationary great part of these two days at 29·60 inches. 17—20. Windy at intervals, with *Cirrostratus,* turbidness, and driving rains. 21. Much wind, with showers: the sky turbid, and streaked with *Cirrostratus,* in a direction from SE towards NW: calm night.

RESULTS.

Wind for the most part Westerly, and moderate.

Barometer: Greatest height .... 30·30 in.
Least .... 29·32 in.
Mean of the period .... 29·860 in.
Thermometer: Greatest height .... 80°
Least............................. 39°
Mean of the period .... 58·60°
Mean of the hygrometer .... 54°
Evaporation .... 2·33 in.
Rain............................. 1·71 in.

The rains of this period, though absorbed by the parched ground as by a sponge, have completely restored vegetation in our meadows, which have resumed, in the space of a few days, a verdure equal to that of spring. Neither the natural nor the artificial indications of this change of weather were very striking: the most considerable being, probably, the sudden increase of temperature in the *nights* previous to the more considerable falls of rain.

1818.		Wind.	Pressure. Max.	Min.	Temp. Max.	Min.	Hygr. at 9a.m.	Rain, &c.
9th Mo. L. Q. Sept.	22	SE	29·70	29·65	69	48	80	
	23	SE	29·65	29·53	61	53		20
	24	SE	29·68	29·60	64	48	75	—
	25	NE	29·65	29·33	65	51	66	1·22
	26	SW	29·66	29·50	61	45	98	20
	27	SE	29·65	29·52	64	54	100	—
	28	SE	29·66	29·59	73	55	85	
	29	SE	29·61	29·41	68	54		—
New M.	30	SE	29·47	29·40	63	51		25
10th Mo. Oct.	1	SE	29·60	29·40	66	45		—
	2	SE	29·65	29·50	68	51		—
	3	SW	29·51	29·40	66	52		90
	4	SW	29·50	29·36	66	47	61	15
	5	W	29·37	29·19	58	41	63	07
	6	W	29·50	29·25	60	32	64	
1st Q.	7	NW	29·79	29·50	59	32	82	
	8	NW	29·88	29·79	57	34	89	
	9	NW	29·85	29·64	60	46	78	
	10	SW	29·64	29·48	63	51	71	13
	11	SW	29·71	29·40	65	46	78	23
	12	SW	29·91	29·71	61	44	65	
	13	SW	29·98	29·86	66	47	71	
Full M.	14	SE	29·99	29·91	70	52	74	—
	15	SE	29·99	29·91	67	53	81	1
	16	SW	30·09	29·99	71	51	71	
	17	NE	30·10	29·92	68	43	79	
	18	Var.	29·99	29·92	63	47	76	
	19	NW	30·06	29·94	63	49	75	
	20	SE	30·20	30·06	61	31	71	
	21	NE	30·20	29·98	56	44	72	
			30·20	29·19	73	31	76	3·36

NOTES.—Ninth Mo. 22. *Cumuli* beneath a canopy of haze, show-
ing as before rain, until evening, when the appearances gave place to
Cumulostratus and red *Cirri*, followed by dispersion of the clouds,
and fall of dew. 23. Much cloud, as yesterday, with wind: rain at
dark. 24. Early morning wet: then fair, with various clouds, threatening rain at intervals. 25. Dew: gray sky, with the lighter modifications: overcast, p. m.: rain, evening and night. 26. Morning, wet:
windy, with *Cumulus*, *Cirrostratus*, &c.: a *Nimbus* in the S, p. m.:
rain after dark. 27. *Cumuli*, with an arch of *Cirrostratus* resting on

their tops in the S: much dew: the dripping shrubs steam in the sunshine, and the breath is visible: *Cirri* in bundles succeeded, at two different elevations, the lower pointing NW and SE, the higher NE and SW: heavy clouds next advanced from the S, the vane being at SE: a shower ensued by inosculation, about noon: drizzling, p. m. 28. Gray morning: a beautiful stratum of small *Cirrocumulus*, between *Cirrostratus* and *Cirrus*, at different considerable elevations: the *Cirrus* proved permanent, and the day was fine. 29. Very high coloured *Cirrocumulus* and *Cirrostratus* at sun-rise: fine. 30. Wet morning.

Tenth Mo. 4. Showers: evening, fine: windy. 5. Clear morning: showery day: wind high. 6. Foggy morning: clear day: a *Stratus* on the marshes at night. 7. Very foggy morning: white frost. 8. Foggy morning: fine, clear day. 9. Very foggy morning: cleared off about nine o'clock, a. m.: day fine. 10. Cloudy. 11. Cloudy, with slight showers: very boisterous night. 12. Morning calm and clear: day fine. 13. Fine day: a very distinct double lunar halo at night. 14. Fine day: clear moonlight. 17. Foggy morning. 21. Clear morning: very fine day.

RESULTS.

Prevailing Wind Southerly.

Barometer:	Greatest height	30·20 in.
	Least	... 29·19 in.
	Mean of the period ...	. 29·698 in.
Thermometer:	Greatest height	73°
	Least	31°
	Mean of the period	55·31°
Mean of the hygrometer ·		76°
Rain		3·36 in.

₊ The whole of the observations, except of the barometer, from the 1st to the 21st of the Tenth month, were made at the laboratory, Stratford.

Coloured Sky at sunrise before rain.

Travelling on the night of the 2d of Tenth month from London to Ipswich, I was struck with the peculiar beauty of the coloured sky, as we entered the latter town, at sunrise. The whole hemisphere was overspread with Cirrus, passing to Cirrostratus and Cirrocumulus, having an arched lowering appearance; the whole dipt in a great variety of tints. The most conspicuous show was however made by dense Cirrostrati to the SE and E, which assuming first a deep blood red passed next through crimson, and a gradation of lighter reds, to orange and then to flame colour. Though the morning was fine, the afternoon and night of the 3d proved very wet in these parts, as at Stratford.

Phenomena observed at Gosport, by Dr. William Burney.—*Thomson's Annals of Philosophy*, vol. xii. p. 368.

Sept. 1. A storm of rain, hail, thunder and lightning from two to half past four afternoon. 5. A fall of rain, amounting to 1·33 in., being as much as had fallen there in the preceding sixteen weeks. 16. A large *Lunar rainbow*, on an extensive Nimbus to the W; the harvest moon being in the E, nearly at full and about 10° above the horizon. 21—23. Strong gales from the S and SW, by day only. 23. From 40 till 55 minutes after five, p.m, a double rainbow to the E, when the sun was within two or three degrees of the W horizon. The interior bow appears to have measured 84° 30' in diameter, and the exterior one being distant on each side 8° 22' 30'', the total extent of the bow on the horizon from N to S, was 101° 15' This measurement of the rainbow, the Doctor observes, is as wide as it can be, within nine minutes, according to the most accurate calculations: the bow was the finest in colours that he remembers to have seen.

25. Two coloured *parhelia*: each of them being at an equal altitude with the Sun, which, at eight, a.m. was 19° 4' 40'', and distant 23° 30' from that luminary; the conspicuous halo in which they were formed was consequently 47° in horizontal diameter: they were caused by an attenuated Cirrostratus cloud brought from the southward.

26. About eight minutes before eight, a.m. three coloured *parhelia* appeared in a coloured halo on a thin Cirrostratus that was passing very slowly in a NW direction. The two parallel with the Sun (at 18° above the horizon) were each 22° 35'; and the third, situate vertically, nearly 28° distant from the centre of his disc. The latter was formed by the intersection of a portion of another halo at the top of the perfect one, and was the largest and most resplendent in prismatic colours of the three. These parhelia were in both cases followed by rain in the afternoon: and the appearance of the Solar halo on the 1st, 6th, 14th, and 22d, and of Lunar halo in the evening of the 17th, were in like manner succeeded by rain, in some instances in less than four hours after.

Oct. 2. At eight, p.m. a very brilliant meteor fell through a space of about 25°; it was of the apparent size of Jupiter, towards which planet it descended from the zenith. 5. At one, a.m. several loud claps of thunder, with lightning, heavy rain, and strong gusts of wind. 7. A *Stratus*, with the first hoar-frost of the season, followed by a parhelion formed on a broad streak of Cirrostratus, a. m.: Cirri then appeared, succeeded by fleecy and dusky Cumuli, and lastly the Cumulostratus, which passed to Nimbus with a short shower.

8. At sun-set, fleecy and dusky Cumuli to the westward passed through orange, dark blue, lake and crimson tints, while the Eastern sky exhibited a rose colour, in an arch of about 35° in height, with a purple base: this magnificent appearance, the Doctor observes, was evidently produced by reflection from a dewy haze descending in that quarter.

9. A single *parhelion* on a Cirrostratus without solar halo, followed by a wet night.

In the valuable observations which I have here abridged, there will be found several points of comparison with my own. On the 1st of Ninth mo. *we* were sensible of a thunder-storm to the SE, preceding the one at Gosport in time by 12 hours—the distance between the two perhaps 100 miles, along the South coast. The 5th was distinguished with us also by heavy rain. On the 7th of Tenth mo. the first *hoar frost* appears to have been noticed, together with the *Stratus* cloud, both at Gosport and Stratford. The remark on the appearance of the Eastern sky during the fall of dew on the 8th, is a confirmation, with the addition of a measurement of the height of the phenomenon, of several notes which I have at different times made upon it. There is likewise a description, by the same observer, of two beautifully coloured parhelia in a preceding number of the Annals, vol. xii. p. 235, to which the reader is referred.

————————

ICE IN THE ATLANTIC.

By Demerara papers of Oct. 24, it appears that Ice islands have been seen as far South as the West Indies; a very extensive one having been observed in the neighbourhood of the Bahamas.

1818.		Wind.	Pressure. Max.	Pressure. Min.	Temp. Max.	Temp. Min.	Hygr. at 9a.m.	Rain, &c.
10th Mo.	L. Q. Oct. 22	SE	30·00	29·93	55	41	68	
	23	E	30 07	30·00	55	45	71	
	24	NE	30·07	30·02	55	44	70	13
	25	SE	30·05	30·01	60	46	73	
	26	SE	30·15	30·05	62	36	70	
	27	SE	30·16	30·14	60	37	72	
	28	SW	30 30	30·16	65	44	75	
New M.	29	SW	30·35	30 30	60	41	79	
	30	SW	30·30	30 05	58	43	72	
	31	SW	30·05	29·89	57	46	81	20
11th Mo.	Nov. 1	NW	29·93	29 87	58	45	75	
	2	SW	29·87	29·70	60	51	77	2
	3	SW	29 70	29·45	61	42	74	
	4	SE	29·45	29·35	56	48	79	8
1st Q.	5	NE	29·35	29 24	60	52	82	1
	6	E	29·60	29·26	58	50	77	
	7	SE	29 87	29 60	51	35	75	3
	8	E	29 94	29·87	52	42	78	
	9	NE	29·98	29·95	52	44	81	
	10	NE	29·95	29·72	50	45	79	86
	11	SE	29·80	29·65	53	41	99	3
Full M.	12	SE	29·65	29·53	50	41	80	6
	13	SE	29·60	29·55	57	46	94	2
	14	S	29·55	29·32	56	43	77	52
	15	W	29·71	29·45	57	41	72	—
	16	SW	29·70	29·44	52	41	96	43
	17	NW	30·05	29·70	47	31	71	
	18	SW	30·07	30·05	47	34	90	
	19	SE	30·07	29·90	50	35	89	
	20	SE	29·90	29·70	45	36	76	
			30·35	29.24	65	31	78	2·39

NOTES.—Tenth Mo. 23. Windy. 24. Cold wind: some rain, evening. 29, a. m. misty. (On the 26th, a little before eight in the evening, I observed from the neighbourhood of Lowestoft, Suffolk, a distinct commencement of *Aurora Borealis* in the north, in white streamers ascending to a considerable elevation, which after a minute or two became converted into a still light: the latter, remaining for an hour or two after, was at length obscured by clouds.)

Eleventh Mo. 1. Foggy morning. 6. Foggy evening. 7. Very fine, a. m. 10. Much wind, with heavy rain, in the night: about ten, p. m. the clouds were passing over rapidly from SE. 11. Small rain, a. m.: gloomy, p. m. 12. The moon at night rose gold coloured, and slightly veiled at intervals by the lighter modifications, with a mixture of haze : on these clouds were afterwards displayed a succession of halos, strongly coloured with green and red. I found by attentive observation that the halo in this instance, together with a corona, which appeared constantly within it, was not formed in the substance of the clouds above-mentioned, but in haze, which was probably situated near the earth ; for the colours of the halo always survived (though faintly) the passing away of the white skreens of cloud by which at intervals they were set off and made conspicuous. 13. A fine day : large *Cumuli* beneath *Cirri*, the latter ranging at night from SE to NW very conspicuous : the moon again rose gold-coloured. 14. Rain, a. m., and again in the fore part of the night, with much wind from the southward. 15. Windy, cloudy, a. m. : lunar halo : rain in the night. 16. Wet morning : fair, p. m., with *Cirrus, Cirrostratus*, and *Cumulus*. 17—20. Fair : the sensible evaporation somewhat reinstated by a brisk wind at the close.

RESULTS.

Prevailing Winds Southerly and Easterly.

Barometer: Greatest height . . . . 30·35 in.
Least . . . . 29·24 in.
Mean of the period . . . . 29·834 in.

Thermometer: Greatest height . . . . 65°
Least . . . . 31°
Mean of the period . . . . 48·75

Mean of the hygrometer . . . . 78°

Evaporation . . . . 1·31 in.
Rain . . . . 2·39 in.

The observations on the thermometer and rain for the former half of the period, and on the hygrometer for the whole, were made at the laboratory.

Observations made at Lowestoft.

I spent nearly the whole of the Tenth month with my family on this, the most easterly point of our island. After a heavy rain at the time of our arrival we had a variety of fine autumnal weather: clear calm days with dew and gossamer—Cumuli, &c. with a brisk air—gray autumnal skies—and once, for three days, a gale at NE with a uniform close canopy of Cumulostratus.

On this occasion I remarked that, at a temperature of 47°, in the middle of the day, the air which blew directly from the sea, and carried with it an abundance of spray in a fine mist from the surf, was yet so dry, just out of this mist, that De Luc's hygrometer stood at 54 degrees, and evaporation proceeded rapidly. But at other times a *southerly* wind, coming along shore, was as moist as usual, giving 87° by the hygrometer. The mere *contact of the sea* in crossing from land to land does not therefore render the lower air so moist as we might suppose.

1818.			Wind.	Pressure.		Temp.		Hygr.	Rain,
				Max.	Min.	Max.	Min.	at 9 a.m.	&c.
11th Mo.	L. Q.	Nov. 21	SE	29·70	29·67	40°	37°	64	
		22	E	29·72	29·47	50	34	65	—
		23	S	29·70	29·47	54	43	92	27
		24	S	30·05	29·70	46	30	100	9
		25	S	30·14	30·05	50	32	82	10
		26	SW	30·32	30·10	54	48	93	13
		27	SW	30·40	30·32	54	48	88	
	New M.	28	SW	30·38	30·30	57	48	76	
		29		30·30	30·20	58	46	75	
		30	S	30·20	30·03	57	46	75	
12th Mo.		Dec. 1	SW	30·03	29·88	46	42	64	20
		2	N	30·00	29·60	48	36	71	—
		3	SW	29·60	29·45	49	35	100	
	1st Q.	4	SW	29·42	29·37	47	40	76	
		5	SE	29·57	29·42	51	33	74	
		6	SE	29·58	29·27	51	40	77	
		7	SW	29·65	29·27	54	44	81	30
		8	SE	30·00	29·65	54	36	88	10
		9	NE	30.10	30.00	46	32	76	40
		10	NE	30·13	30·08	45	32	88	
		11	NE	30·14	30·08	43	30	71	
	Full M.	12	NW	30·12	30 07	42	37	80	2
		13	NE	30·18	30·10	41	31	69	
		14	NE	30 20	30·15	43	33	87	
		15	NE	30·15	29·96	40	23	78	
		16	NW	30·17	30·00	32	16	78	
		17	NW	30 12	29·82	28	18	79	
		18	Var.	30·00	29·70	39	25	84	13
		19	SW	30·25	30·00	43	29	86	
	L. Q.	20	SW	30·10	29·90	47	43	73	1
				30·40	29 27	58	16	79	1·75

NOTES.—Eleventh Mo. 21. Fair: cloudy, with a strong breeze.
22. *Cirri* tending to *Nimbus*, a. m.: *Cumulus* beneath *Cirrostratus*:
little wind. 23. Wet, gloomy, a. m. fair, p. m.: at sun-set, rose-
coloured *Cirri*, with orange in the twilight. 24. Foggy morning: the
dew frozen on the grass: the vane at SW. 25. A very dense *Cirrostratus*, a. m. forming a mist, which did not reach to the tops of the
trees: a solar halo at 11: more clear in the evening: rain in the
night: the wind SW to SE. 26. Wet, windy morning: fair and
cloudy, p. m. and night. 27, 28. Cloudy, 29. Gloomy, fair, calm.
30. A breeze, with light clouds: fine, p. m. with *Cumuli*.
    Twelfth Mo. 1. Rain in the night. 2. The vane at N, a. m. but

in the night the wind came to SW, blowing fresh, with a little rain.
3. Vane at S in the morning, with much wind: cloudy. 4. Fair,
windy, cloudy. 7 A drizzling rain through the day. 6. Hoar frost,
8. Showery, a.m. 9. Wet. 10—20. Chiefly fair and cloudy: at
intervals, fine, with the wind moderate: very white hoar frost on some
of the latter mornings, with rime to the tops of the trees. Large lunar
coronæ were frequent in the evenings, and lunar halo occurred more
than once; but the dates were not noted.

RESULTS.

Winds Variable.

Barometer: Greatest height .... 30·40 in.
Least .... 29·27 in.
Mean of the period .... 29·925 in.
Thermometer: Greatest height .... 58°
Least........................... 16°
Mean of the period .... 41·20°
Mean of the hygrometer .... 79°
Evaporation...................... 0·35 in.
Rain............................. 1·75 in.

The few nocturnal frosts that occurred in the present season, up to
the middle of the 12th month, were so slight as to permit the *Nastur-
tiums* (the tenderest of our autumnal garden flowers) to continue to
vegetate: other indications of the mildness of the season were equally
striking. I observed a horse-chesnut with tufts of new leaves and
blossoms put forth from the ends of the branches all over the tree; but
the severe nights, and some frost by day, since the 15th, have put a
seasonable stop to vegetation. The temperature of the latter half of
the period, and the hygrometer throughout, were noted at the laboratory.

STATE OF VEGETATION.

The extraordinary growth of mushrooms, this autumn, makes a kind of second
harvest for the industrious poor in most parts of the island, some of whom have
gathered from three to five pecks daily.

John Foster, Esq. of Newton, near Carlisle, has a single tree in his orchard
which has, this year, produced *ten thousand* apples.—*Pub Ledger, Sept.* 28.

An apple-tree in the garden of D. Sutton, Esq. at Kensington, opposite Holland
House, is now covered with a second full crop of apples; and there are several
others in the same garden which have also had a second produce this year, though
not in such abundance.—*Pub. Ledger, Nov.* 12.

STOCKHOLM, *Nov.* 17.—The uncommonly serene and mild autumnal weather
still continues, and now supplies us in abundance with garden produce, of which
we were deprived during the summer by the great drought. To be without fire
in the stoves, and to have the meadows covered with verdure instead of snow, is
a strange phenomenon here in the month of November.

A thorn-tree growing in the lawn at Shugborough, the seat of Lord Anson, is
now in full blossom, and the whole tree presents a May-like appearance.—*Dec.* 1.

There was lately in the garden of G. Dickson, Esq. of Cousland, Berwickshire,
(formerly the garden of the old Priory of Cousland,) a tulip-tree in full blossom,
which is but the second time it was ever known to be in that state. The last time
this tree was in blossom was in the year 1720, being 98 years ago.—*Dec.* 2.

Among the many instances of the extraordinary temperature of the present
season, may be mentioned that from the Garden of Thomas Newton, Esq. of
Clapham Common, green peas of full growth and flavour were gathered a day or
two ago, and the haulm still remains in full blossom —*Pub. Ledger, Dec.* 18.

K

1819.	Wind.	Pressure. Max.	Min.	Temp. Max.	Min.	Hygr. at 9 a.m.	Rain, &c.
12th Mo. Dec. 21	W	30·50	30·15	48°	24°	76	—
22	Var.	30·50	30·40	28	23	93	
23	NW	30·40	30·27	35	23	96	
24	Var.	30·27	30·10	28	22	87	
25	NE	30·10	29·82	38	26	88	
26	SE	30·05	29·80	35	30	72	
New M. 27	SE	30·47	30·05	39	32	78	
28	NE	30·60	30·47	41	30	69	
29	NW	30·58	30·47	38	22	81	
30	W	30·45	30·38	34	23	84	
31	NW	30·48	30·38	38	26	94	
1st Mo. 1819. Jan. 1	NW	30·50	30·45	35	26	90	
2	W	30·45	30·30	38	30	90	
1st Q. 3	SE	30·30	30·15	42	24	75	
4	SE	30·17	30·11	40	22	95	
5	N	30·18	30·15	40	26	97	
6	Var.	30·15	29·76	45	32	95	
7	S	29·86	29·60	46	33	80	—
8	W	29·86	29·40	47	33	81	25
9	SW	29·87	29·43	51	34	72	36
10	SW	29·80	29·48	49	40	75	15
Full M. 11	SW	29·60	29·48	46	34	98	11
12	SW	30·13	29·60	50	33	82	
13	SW	30·12	29·95	49	35	75	—
14	SW	30·00	29·85	53	37	88	—
15	SW	30·20	29·77	50	33	79	16
16	NW	30·30	30·00	50	34	61	—
17	SW	30·00	29·25	50	35	70	19
18	NW	29·80	29·28	43	32	71	
		30·60	29·25	53	22	82	1.22

NOTES.—Twelfth Mo. 21. Much wind about three, a. m. with a little rain: a very fine day ensued: *Cirrocumulus*, with bright sunshine. 22. White frost: foggy, a. m.: clearer, p. m. with *Cirrus*: rime on the trees. 23. White frost: rime to the tree tops: misty, a. m.: sun very bright at noon: much fog to the south. 24. Very foggy: rime still on the trees. 30. White frost. 31. A very fine day.

1819.—First Mo. 1. Very foggy, with the addition of obscurity from smoke. 3. Fair: rather overcast sky. 4. Much rime on the trees: rather misty air. 5. Somewhat misty: the melted rime forming puddles under the trees. 6. Fine day : at night, small portions of cloud were observed to pass swiftly under the moon. 7 The sun-rise was attended with a veil of *Cirrus* clouds passing to *Cirrostratus*, very red and lowering : about noon *Cumuli* and other clouds, with a gale and showers. 8. Fine day : night windy, with some rain. 9. Hazy, a. m. with *Cirrostratus* and wind : heavy showers, p. m. : very clear night. 10. Overcast soon after sun-rise, with wind : the fore part of the night very stormy. 11. A wet squall this morning : fair day, with *Cirrus* and wind. 12. Fair morning, with slight hoar frost : the gale has subsided. 13. Slight hoar frost : very fine day : at evening, windy again. 14, 15. Windy, with some showers. 16. A fine drying wind, a. m. with *Cirrus* and *Cirrostratus* in delicate striæ : also transient *Cirrocumulus* at a great elevation : a stormy night followed. 17. A very tempestuous day : the rain ceasing for a while, a. m. I observed *Cirrostratus* around large *Cumuli*, rising and separating, as if the shower had been produced by their previous inosculation : much wind in the night. 18. A fine, drying day, with the wind more moderate, and an overcast sky.

RESULTS.

Winds Variable and gentle, with fogs in the fore part ; in the latter, strong South-west gales, with rain.

Barometer :	Greatest height	30·60 in.
	Least....	29 25 in.
	Mean of the period . ..	30·068 in.
Thermometer :	Greatest height	53°
	Least	22°
	Mean of the period	35·86°
Mean of the hygrometer .. .·		82°
Rain		1·22 in.
Evaporation, about. ..		0·50 in.

Extract of a Letter from Petersburgh.

Petersburgh, First Month 5, 1819 —The weather here is most extraordinary for this climate. It is now a thaw ; a circumstance not remembered here at this date by any person living. Great quantities of the frozen meat brought from the interior are spoiling.—W. A.

Barcelona, Jan. 2.—For these ten days past we have experienced strong gales from the Eastward.

Dumfries, Jan. 15.—It has blown a very heavy gale from SW to W for six days past, but is now more moderate.

Port Glasgow, Jan. 17.—The weather still continues very tempestuous.

Penzance, Jan. 17.—A gale from W to WNW the whole day and night.

Portsmouth, Jan. 18.—No arrival or sailing. The whole of the day it has blown very strong from WNW.

Ostend, Jan. 16.—It blew nearly a hurricane during the night, from NW to N. —*Shipping List.*

1819.			Wind.	Pressure. Max.	Min.	Temp. Max.	Min.	Hygr. at 9a.m.	Rain, &c.
1st Mo.	L. Q.	Jan. 19	NW	29·80	29·45	44	33	66	19
		20	W	29·57	29·28	41	33	70	9
		21	W	29·52	29·20	40	32	66	
		22	SW	29·72	29·20	49	34	89	—
		23	SW	29·72	29·60	45	33	75	
		24	SE	29·60	29·20	47	36	90	
		25	SE	29·43	29·10	47	39	74	26
	New M.	26	SE	29·45	29·35	47	35	83	
		27	E	29·44	29·35	49	38	96	
		28	SE	29·52	29·30	50	31	85	
		29	E	29·52	29·35	50	30	97	
		30	E	29·40	29·28	42	33	88	43
		31	NW	29·62	29·40	42	25	78	—
2d Mo.		Feb. 1	SW	29·60	29·49	38	28	85	—
	1st Qr.	2	NW	29·73	29·49	38	18	90	—
		3	SE	29·73	29·42	41	25	88	—
		4	NW	29·80	29·47	43	31	75	29
		5	E	29·78	29·46	47	35	95	12
		6	W	29·61	29·39	50	36	85	17
		7	NW	29·65	29·35	47	33	68	
		8	W	30·00	29·65	46	34	72	
		9	SW	29·95	29·63	51	44	85	37
	Full M.	10	NW	30·05	29·58	49	36	69	
		11	W	30·03	29·79	51	34	72	
		12	W	29·79	29·50	48	33	69	—
		13	NW	29·90	29·62	47	27	64	14
		14	NW	30·12	29·90	47	25	64	
		15	SW	30·08	29·70	47	34	65	
		16	S	29·70	29·37	51	42	69	10
				30·12	29·10	51	18	78	2·16

NOTES.—First Month. 19. *Cirrus* with *Cirrocumulus*, in lines stretching N and S. rain in the night. 20. A very fine day: *Cirri,* p. m. rain and wind in the night. 21. Slight hoar-frost: *Cirrocumulus.* 22. Fair day: rain and wind, evening. 23. Very fine. 24. Fair: strong breeze: cloudy. 25. Rain, a. m. 26. Fair day: large *Cumuli* appeared, passing to *Cumulostratus* with plumose *Cirri* above: at evening there were indications of the *Stratus.* The *Nimbus* has been frequent during the past week: the wind generally moderate in the day, and strong the fore part of the night. 29. Morning rather overcast: day fine, with the lighter modifications ranging (as frequently or lat) in lines N and S. About ten, a. m. in going to London, I observed a solar halo of large diameter, imperfect in its superior and inferior part, except a trace at the vertex, but exhibiting, in the points directly N and S of the sun, *two parhelia*, which continued with a faint variable brightness for about twenty minutes. 30. Wet morning: drizzling most part of the day: wind SE, and then NE. 31. Overcast: rained a little, a. m.

Second Mo. 1. Hoar frost, with *Cirri* in the sky, pointing upwards from a base: drizzling rain at night. 2. *Snow* (for the first time this season) continuing most part of the forenoon from sun-rise: then, brilliant sunshine, and frost at night, with the Thermometer nearly as low as the minimum of the present winter. 3. Rather misty and overcast, a. m.: wet evening. 4. Cloudy: fair, a. m.: showers, p. m. 5. Misty, drizzling. 6. Very fine, with *Cumuli*, &c. a. m. in the afternoon, a squall of wind, with a few drops: in the night a gale followed by rain. 7. Very fine. 8. Fair, with *Cirrostratus* in parallel bars here and there, under uniform haze: at night a lunar halo, very large and colourless. 9. Wet day: stormy night. 10. Early this morning it was very tempestuous; but the day was fine, with *Cumuli* carried by a moderate gale, and *Cirri* scattered like loose hay above: at night, with *Cirrostratus*, a succession of small, ill-formed, but highly coloured halos. 11. Fine, with *Cumulus*, *Cirrostratus*, and wind. 12. Fine morning, then showers (in London attended with hail), and much wind at night. 13. Fine morning: *Cumulus* capped with *Cirrostratus*: *Nimbi*, p. m. with a transient rainbow. 14. Slight hoar frost: fine, with *Cumulostratus*, and a breeze. 15. Fine: the ground was frozen this morning, and *Cirrocumulus* at the same time above. 16. Overcast morning: wet and windy, p. m. and night.

RESULTS.

Winds Westerly, except a week about the New Moon, when they were East and South-East.

Barometer:	Greatest height	30·12 in.
	Least	29·10 in.
	Mean of the period	29·522 in.
Thermometer:	Greatest height.	51°
	Least	18°
	Mean of the period.	39·31°
Mean of the Hygrometer.		78°
	Evaporation.	0·65 in.
	Rain	2·16 in.

Falmouth, Jan. 24.—A strong gale all last night from the SW, with heavy squalls of hail and rain. 25. A heavy gale the whole day from SW to WSW.

Hull, Jan. 25.—A heavy gale from SE. 26. More moderate, from WSW.

Portsmouth, Jan. 25.—A very heavy gale the whole day, with tremendous gusts from the Southward and SSW.

Deal, Jan. 25.—Two, p. m. Wind S, blows hard and a tremendous sea.

ST. UBES, *Jan.* 28.—The weather changed between the 23d and 24th at night, and ever since has continued extremely stormy, with rain and hail, thunder and lightning. A severe shock of an earthquake was felt, and afterwards another not so violent.—*Shipping List.*

Milford, Feb. 15.—All the vessels bound round land sailed yesterday with the wind at NNW: the wind shifted during the night to the Southward and they are put back. 16. It blows hard from the SSW.

	1819.	Wind.	Pressure. Max.	Min.	Temp. Max.	Min.	Hygr. at 9 a.m.	Rain, &c.
2d Mo.	L. Q. Feb. 17	SW	29·54	29·35	52°	39°	64	30
	18	NW	29·62	29·30	49	42	90	9
	19	SW	29·85	29·15	51	31	80	—
	20	SW	29·85	28·90	52	36	69	23
	21	N	29·90	28·90	49	38	65	13
	22	Var.	29·97	29·33	45	34	67	35
	23	SE	29·53	29·33	45	28	76	27
	New M. 24	NW	29·75	29·40	45	23	67	8
	25	NW	29·80	29·52	41	27	75	
	26	SW	29·52	29·31	39	30	68	6
	27	Var.	29·31	29·25	45	37	89	7
	28	SE	29·26	29·14	40	34	78	—
3d Mo.	March 1	NE	29·30	29·17	41	34	87	54
	2	E	29·61	29·30	44	35	94	11
	1st Q. 3	NE	29·90	29·61	39	34	65	
	4	NE	29·98	29·87	45	34	62	7
	5	NE	29·97	29·83	47	40	83	
	6	NE	30·10	29·90	50	36	85	
	7	NE	30·16	30·10	46	37	66	
	8	NE	30·12	30·08	46	30	81	
	9	SE	30·13	30·10	47	27	67	
	10	NW	30·10	30·04	46	34	82	
	Full M. 11	NW	30·10	30 07	51	42	74	
	12	NW	30·31	30·16	51	41	65	
	13	NW	30·34	30·29	48	40	61	
	14	Var.	30 30	30·14	48	24	59	
	15	NE	30·14	29·99	57	34	67	—
	16	W	30·13	29·98	59	40	85	—
	17	NW	30·25	30·13	46	27	60	—
	18	NW	30·15	29·45	49	35	63	—
			30·34	28·90	59	23	73	2·30

NOTES.—Second Mo. 17. A fair day, with *Cumulostratus:* rain by night. 18. Fine and spring-like: *Cumuli* capped with *Cirrostratus,* a. m.: very stormy night. 19. After a squall in the morning, a very fine day, with large *Cumuli* and *Nimbi:* a full bright rainbow at three, p. m.: the wind settled by evening. 20. Hoar frost: very fine morning, p. m.: large ramified *Cirrus* mixed with *Cirrocumulus* at a great height: *Nimbi:* some violent wet squalls in the night from the southward. 21. Large *Cumuli,* and much wind: showers. 22. Wind shifted to N: cloudy morning: *Cumulostrati* by inosculation. 23. Wind and rain: of the latter, 0·35 in. between six and nine, and 0·27 in. more by noon: afternoon, a gale, with much cloud: evening more settled. 24. Fine morning : at noon, lofty large *Cumulostrati*, with bright sun : in the course of the afternoon, an obscurity, like the crown of the *Nimbus*, came down upon these clouds ; and a considerable fall of *snow* took place before dark, with wind. 25. Snowy morning : the hills white with snow ; which soon vanished before a bright sun, p. m. 26. *Cirrocumulus* appeared above, while the ground and water were frozen : about half-past ten, a faint, but large *solar halo*, which continued till near eleven, when obscurity came on from the southward, followed by drizzling rain, p. m. 27. Overcast morning : rain in the night. 28. Cloudy : some rain.

Third Mo. 1. Snow and sleet, a. m. : a wet day. 2. Wet morning : cloudy, drizzling day. 3. A moderate easterly gale, with much cloud : a gleam of sunshine, p. m.———18. There has been scarcely any rain since the 4th; the sky mostly gray, with light clouds; at times overcast, or filled with *Cumulostratus*: the wind northerly, breezes, and the air drying; so that the roads at the close of the period, notwithstanding some very light showers of late, remained considerably covered with dust. The diverging bars of light and shadow, produced by the sun's rays passing through the interstices of clouds, have been several times exhibited within these two days.

RESULTS.

Winds for the most part Northerly.

Barometer: Greatest height .. .. 30·34 in.
Least .... 28·90 in.
Mean of the period.... ...... 29·768 in.
Thermometer: Greatest height .. .. 59°
Least ... . 23°
Mean of the period.......... 40·78°
Mean of the hygrometer............... 73°
Evaporation. .... 0·66 in.
Rain...................... 2·30 in.

Weymouth, Feb. 21. It blew very hard last night from the SW.

Falmouth, Feb. 21. It has blown a heavy gale all day, WNW.

Torbay, Feb. 22. All the vessels sailed this morning with the wind at N, after having experienced a tremendous gale at W, all yesterday.

Deal, Feb. 23. It has blown very strong the whole morning from SSW : at three p. m. suddenly shifted to due N : five, p. m. blows very strong.

Falmouth, Feb. 23. It has blown very heavily all day from the N & NW. 24. It has blown very heavily all day and continues unabated, from the N.

March 3. The wind has been very high all day, and is much increased since dark, from the Eastward.—*Shipping List.*

1819.		Wind.	Pressure.		Temp.		Hygr. at 9 a.m.	Rain, &c.
			Max.	Min.	Max.	Min.		
3d Mo. L. Q. Mar.	19	Var.	29·50	29·21	51°	38°	90	16
	20	NW	29·75	29·50	46	37	65	—
	21	NW	29·80	29·75	48	32	59	
	22	NW	29·80	29·70	53	35	63	
	23	S	29·70	29·49	51	42	61	—
	24	SW	29·62	29·49	58	44	77	—
New M.	25	SW	29·85	29·64	55	37	67	—
	26	W	29·90	29·85	54	40	68	—
	27	SW	29·90	29·73	55	46	59	59
	28	SW	29·77	29·68	54	46	85	—
	29	SW	29·96	29·67	57	43	67	
	30	SW	30·07	29·96	58	50	69	8
	31	SW	30·18	30·07	59	47	69	
4th Mo. April	1	W	30·20	30·15	62	48	61	
1st Q.	2	W	30·15	30·05	68	36	66	
	3	N	30·17	29·99	68	43	67	
	4	E	30·17	30·07	60	38	61	
	5	NE	30·06	29·94	60	29	61	
	6	SE	29·94	29·60	54	43	60	
	7	E	29·80	29·62	66	46	60	10
	8	NW	30·00	29·80	58	40	74	33
	9	N	30·07	29·95	61	34	67	
Full M.	10	W	29·95	29·35	64	40	66	
	11	SW	29·35	29·27	60	37	66	14
	12	NE	29·30	29·12	50	43	74	35
	13	Var.	29·48	29·12	58	40	85	3
	14	SW	29·48	29·30	6C	42	74	—
	15	SW	29·40	29·03	60	45	76	—
	16	S	29·40	29·03	58	44	68	45
			30.20	29·03	68	29	68	2·23

NOTES.—Third Mo.   19. A moderate gale at SW in the early morning, with much cloud carried by the wind.   About ten, the wind changing suddenly to NW, the whole mass of cloud to the southward became an immense *Nimbus*, the base reaching from the SW to the NE, with a lighter sky visible beyond: at the same time precipitation was going on overhead, and we had soon a smart shower mingled with hail: the whole ended in a uniform veil of *Cirrostratus*, and at night we had the SW wind again pretty strong.   20. The wind changed again to NW, a. m. with much cloud, and some drops of rain.   21. Fine day: a smart breeze from NW. 22. Fine day. 23. A trifling shower. 24. Wet, windy morning: fair day. 25. A shower with hail at mid-day: a large *Nimbus* passed, and a distant peal of thunder was heard to the NW. 26. Chiefly *Cumulostratus:* a very little rain, p. m. 27. Windy, with much cloud, and two or three showers. 28. Cloudy: a gale through the day. 29. Cloudy. 30. A rainbow at nine, a. m.: squally, with showers: the bow again twice about three, p. m. 31. Cloudy: some drops of rain.

Fourth Mo. 2. A lunar halo at night, of large diameter, and colourless: it was sensibly elliptical, the longer diameter being the perpendicular; it continued two or three hours. 3. Large *Cirri*, with *Cumuli:* much dew: very fine day. 4. *Cumulostratus.* 5. Fine morning: the hoar frost remained at seven, a. m. on some tufts of *Saxifraga cœspitosa*, &c. (as heretofore noticed) long after it had disappeared elsewhere in my garden; proving that the warmth which melted the ice came in this instance chiefly from the earth, and was here intercepted by a bad conductor. 6. Large plumose *Cirri*, with *Cirrostratus*, a. m. 7. The maximum of temperature for the past 24 hours occurred at nine this morning: thunder-clouds ensued, which soon passed to a quiescent mixture of different modifications, and rain came on at evening. 8. Much *Cirrostratus*, with pretty heavy rain, p. m.: at evening the wind changed to NW, with a rainbow and a turbid mixture of different clouds. 9. Fine, with *Cumulostratus:* wind N, p. m. 11. The clouds this evening were tinged with a strong lake colour, on the bases of *Cumulostrati*, beneath *Cirrus:* some rain attended. 12. Wet, most of the day. 13. Rain, a. m. 14. *Cumulostratus:* in the evening streaks of *Cirrus* from SW to NE, followed by wind and rain. 15. Clouds followed by rain in the night, as before. 16. After a fine day with clouds, rain in the early morning.

RESULTS.

Winds chiefly Westerly.

Barometer: Greatest height .... 30·20 in.
Least .,.................... 29·03 in.
Mean of the period........... 29·738 in.

Thermometer: Greatest height.............. 68°
Least...................... 29°
Mean of the period .... 49·20°

Mean of the hygrometer .... 68°.

Rain .... 2·23 in.
Evaporation .... 1·32 in.

1819.	Wind.	Pressure. Max.	Min.	Temp. Max.	Min.	Hygr. at 9 a.m.	Rain, &c.
4th Mo. L. Qr. April 17	S	29·57	29·40	55	43	75	15
18	SW	29 76	29·57	58	36	62	4
19	W	29·76	29·72	57	47	64	44
20	SW	29·72	29·59	61	46	82	—
21	SW	29·79	29·59	59	42	72	—
22	NW	29·85	29·74	49	42	76	
23	E	29·74	29·51	51	46	90	15
New M. 24	E	29·65	29·51	52	45	100	25
25	NE	30·08	29·65	50	34	76	2
26	E	30·16	30·08	52	32	68	
27	E	30·16	—	56	25	62	
28	SE	—	30·05	60	31	60	
29	E	30·05	29·80	59	28	60	
30	SE	29·80	29·75	60	28	60	
5th Mo. May 1	SW	29·75	29·68	66	33	57	
1st Q. 2	SE	29·68	29·57	69	48	52	
3	SE	29·57	29·45	71	49	58	—
4	SE	29·50	29·45	69	50	87	68
5	SE	29·88	29·50	65	44	69	
6	SW	30·00	29·88	64	39	62	
7	SE	30·00	29·98	66	49	67	
8	E	30·17	29·98	69	44	63	
Full M. 9	SE	30·14	30·06	73	46	59	18
10	NW	30·14	30·10	69	53	65	
11	W	30·10	30·04	65	54	67	
12	NW	30·10	30·04	67	53	61	
13	NW	30·10	30 00	64	41	60	
14	NW	30·04	30·02	64	40	58	
15	N	—	—	67	42	59	
		30·17	29·40	73	25	67	1·91

NOTES.—Fourth Mo. 17. Much wind in gusts, a. m.: the clouds large, and carried at a great elevation: the first swallows appeared: wet squalls, p. m.: some lightning about ten, and a gale through the night. 18. *Cumulus*, with the lighter modifications above, followed by *Nimbi* and wind: hail in a shower about three, p. m.: the rainbow twice: calm at evening. 19—22. Mostly cloudy, with *Cumulostratus*: the cuckoo was heard in this interval. 23. Gloomy sky, with much *Cirrostratus* at evening: rain in the night. 24. Drizzling, a. m.: wet and windy, p. m. 25. A gale, with much cloud in the morning: fair,

p. m. 26. Fair, with heavy *Cumulostratus*. 27, 28 Hoar frost: clear, fine days, with *Cumulus*, *Cumulostratus*, and *Cirrus*. The dark part of the moon's disk, which has been scarcely discernible this winter, is again plainly visible in the evening, as she follows the sun. 29. Hoar frost: fine day, with *Cirrus* and breezes.

Fifth Mo. 1. Fine: much *Cirrocumulus*, mixed with *Cirrus*: the wind a breeze. The gardens have suffered a little by the late frosts, more especially the gooseberry bushes, which have cast a part of their crop. 2. A superior westerly current appeared, a. m. carrying flocks of *Cirrocumulus*: between this and the SE wind below were large plumose *Cirri*, on one of which appeared a trace of a solar halo. These clouds increasing, with haze and *Cumuli* intermixed, the character of the sky became electrical: there was a lunar corona and a small bright halo at night, with some lightning to the SW. 3. Clouds grouped with an electric appearance as yesterday. 4. Obscurity and a little rain, a. m.: heavy showers, p. m.: rain in the night. 5. Fine, with *Cumulostratus*. 6. The same: *Cirrus* and *Cumulus* appeared: the smoke was attracted by the clouds, and a few drops of rain fell by inosculation among the latter, p. m. 7, 8. Clouds various, and mixed with haze: on the latter night, a very luminous large corona round the moon. 9. Large plumose *Cirri*, followed by obscurity and *Cirrocumulus*, with an electrical character: a fine shower at evening: rain in the night. 10—15. Fair, with the lighter modifications, and breezes.

RESULTS.

Winds Easterly in the middle, Westerly in the beginning and end of the period.

Barometer: Greatest height .... 30·17 in.
Least .... 29·40 in.
Mean of the period .... 29·83 in.
Thermometer: Greatest height .... 73°
Least .... 25°
Mean of the period .... 51·67°
Mean of the hygrometer .... 67°
Evaporation .... 2 in.
Rain.................... 1·91 in.

METEOR.

A very remarkable Meteor was seen at Aberdeen, May 5th. At about *half-past twelve in the day*, it appeared at an altitude of nearly 36°, having the form of a ball of fire with a short tail, darting towards the earth. The atmosphere was uncommonly clear, with bright sunshine and no clouds. In about five minutes it exploded with a considerable noise, leaving a small white cloud of smoke. It was seen in many parts of the country. *Journal of the Royal Inst.* From a Scotch Paper.

	1819.		Wind.	Pressure. Max.	Min	Temp. Max.	Min.	Hygr. at 9 a.m.	Rain, &c.
5th Mo.	L. Q. May	16	SE	30 08	30·02	72°	37°	59	
		17	NE	30·02	29·86	75	40	59	
		18	SW	29·86	29·70	69	50	60	48
		19	NE	29 70	29 63	65	54	86	24
		20	SW	29·63	29 59	63	51	71	40
		21	SE	29 66	29·49	62	42	76	4
		22	SE	29·90	29·66	70	39	70	
		23	E	29 91	29 90	77	50	65	27
	New M.	24	E	29·94	29·91	63	49	72	4
		25	NE	29·91	29·90	65	47	72	
		26	NE	29·92	29 91	60	40	70	
		27	NE	29·91	29·87	63	42	63	
		28	NE	29·93	29·89	60	33	60	—
		29	NE	30·02	29·93	60	31	60	
		30	NW	30·09	30·02	64	44	55	9
		31	SW	30·13	30 09	64	46	62	5
6th Mo.	1st Q. June	1	SW	30.10	30 09	70	53	64	
		2	SW	30·10	30·06	70	47	67	
		3	SW	30·06	30 00	72	56	61	
		4	SW	30 06	29·95	72	45	71·	9
		5	SW	30·10	30.07	72	45	63	
		6	SW	30 07	29·70	75	53	59	1
		7	SW	29·72	29 68	66	47	65	
	Full M.	8	SE	29·66	29·63	70	46	62	1
		9	S	29·72	29·67	75	51	62	
		10	SW	29 97	29·72	72	46	59	25
		11	W	30·04	29·97	70	47	63	
		12	SW	30·09	30·04	70	37	60	2
		13	SE	30 10	30·04	70	48	59	
	L. Q.	14	SW	30·04	29·85	71	49	61	3
				30·13	29 49	77	31	64	2·02

NOTES.—Fifth Mo. 16, 17. Fine. 18—21. Rainy: hail in the showers on the latter day': *Cumulus, Nimbus, Cirrus.* 22. *Cirrostratus:* lightning at night to the north-west. 23. Fine day: rain in the night. 24, 26. Cloudy. 28, 29. Frosty mornings. The potatoes suffered considerably in their growing tops, the temperature having undoubtedly been lower on the ground than at the height of the thermometer. 30. Showers.

Sixth Mò. 1. Cloudy. 2, 3. Fine. 4. Showery. 10. *Nimbi,* very large and distinct: a thunder shower about three, p.m. and a brilliant rainbow in the evening. 11. Fine. 12. Showers. 13. Fine. 14. Cloudy.

RESULTS.

Winds Easterly in the fore part, Westerly in the latter part of the period.

Barometer : Greatest height .... 30·13 in.

Least .... 29·49 in.

Mean of the period .... 29·899 in.

Thermometer : Greatest height................ 77°

Least .... 31°

Mean of the period .... 56·90°

Mean of the hygrometer .... 64°

Rain at Stratford .... 2·02 in.

Rain at Tottenham .... 2·29 in.

The above observations, two or three incidental articles excepted, are extracted from the register kept at the Laboratory.

———

A letter from *Palermo,* of the 4th of March, states that during the preceding fourteen days, the weather had been dreadful, and that three shocks of earthquake had occurred. Much mischief was done by the shocks in the southern part of the island (of Sicily), churches being thrown down and villages destroyed.—*Journal of the R. Institution.*

6th Mo.	1819.		Wind.	Pressure.		Temp.		Hygr. at 9a.m.	Rain, &c.
				Max.	Min.	Max.	Min.		
	June	15	SW	29·97	29·83	63	42	65	33
		16	NW	30·12	29·97	62	45	61	2
		17	NW	30·13	30·07	72	52	63	1
		18	NW	30·14	30·05	67	52	82	45
		19	NW	30·19	30·14	72	47	66	
		20	NW	30·19	30·18	72	45	61	
		21	NW	30·18	30·09	78	49	59	
New M.		22	NW	30·09	30·08	68	55	59	
		23	NW	30·08	29·94	72	52	59	
		24	SW	29·94	29·81	63	56	66	5
		25	SW	29·93	29·72	66	55	75	—
		26	S	29·74	29·69	67	48	69	—
		27	Var.	29·74	29·73	65	50	70	30
		28	NW	29·86	29·73	60	45	64	23
		29	NW	29·86	29·79	70	54	63	
1st Q.		30	W	29·86	29·77	72	46	67	
				30·19	29·69	78	42	65	1.39

NOTES.—Sixth Mo. 15. Rain. 16, 17. Cloudy. 18. Wet, a. m.
19—23. Fine days. 24. Morning, overcast: drizzling rain. 25, 26.
Cloudy. 27. Frequent heavy showers: at sun-set, strong shadows
projected from behind a *Cumulus* or two, amidst a wild-looking sky:
the clouds opposite the setting sun tinged to a very fine yellow. 28.
Cloudy morning: a confused sky, exhibiting a mixture of *Cirrus*,
Cirrocumulus, *Cumulus*, &c.: about eleven, a. m. some hail ushered
in a thunder shower: the lightning was vivid at Stratford, and the
thermometer fell 8° during the shower. Other thunder showers
succeeded, till one, p. m. when there was a very heavy storm of large
hailstones, which nearly covered the ground. This was attended
with vivid lightning to the southward, and thunder at a small distance.
29. Overcast. 30. Fine.

RESULTS (of the half period).

Winds Westerly, and chiefly North-west.

Barometer: Greatest height............... 30·19 in.
               Least.... .... 29·69 in.
               Mean height .... 29·955 in.
Thermometer: Greatest height.............. 78°
               Least............ .... 42°
               Mean height ... . 58·17°
        Rain at Stratford.......·........ .. 1·39 in.
        Rain at Tottenham ....,- .... 1·63 in.
        Mean of the hygrometer .... 65°

———————

The present Table, which contains only the half of the usual Lunar period, brings up the observations to the close of the Sixth month, 1819. Having continued the publication of them by Lunar periods in *Thomson's Annals* up to the time at which it became necessary to insert the last of the series in the present work, I determined to adopt, in future, the term of the Calendar month for their periodical appearance : considering that their comparison with those of other observers would thus be greatly facilitated, while my own object of tracing the natural periods of the different atmospherical changes, might be secured by a due arrangement of the results and averages.

OF THE

CLIMATE OF LONDON.

OF THE TEMPERATURE.

TEMPERATURE is that which constitutes the most obvious difference between climate and climate, and on which the variety of the phenomena which are exhibited by each principally depends. I shall therefore take it up here, though standing third in the tables, as the fittest introduction to the study of the whole subject : and as a comparison of the results obtained by different observers, whether for the same or for different periods of years, in the same climate, tends mutually to establish or correct their respective conclusions, I shall freely avail myself of the helps I find before me in this respect.

Of Mean Temperature in general.

To mention the differing warmths of day and night, or of the different months of the year, is simply to appeal to the test of feeling. But feeling informs us on these subjects only by a vague comparison with past sensations, the memory of which, when they have been some time past, is very imperfect. To confirm or to correct our judgment, as to the comparative warmth or coldness of different days or seasons, and still more to be able to compare climates together, with accuracy, we must be accustomed to the use of the Thermometer.

M

If we note the degrees indicated by this instrument
when the heat of the day, by this evidence, is at the
maximum, and again when it is at the *minimum*, and
adding them together divide the sum by two, we shall
have the *medium* temperature of that day, a standard
by which we may judge of the temperature of another
day obtained in like manner, and pronounce it warmer
or colder. This standard would be more accurate,
were the temperature noted at every hour, and the
sum total divided by twenty-four. Although this process be seldom effected for the day, an analogous one
is commonly performed for the month, when, taking
the medium temperatures of the several days, we sum
them up and divide by the number of days thus noted:
the result is called the *mean* temperature *of the month*;
it is a standard for comparing the days of that month
with each other: these monthly means summed up
and divided by twelve, give the *mean of the year*:
which if constructed from a sufficient number of observations, carried through all the seasons, affords a
criterion for judging of the temperatures of the several
months of that year. A long average of these yearly
means, gives a result so nearly approaching to uni-
formity in the hands of different observers, that it may
be used as a general standard of comparison for the
temperature of the day, month, or year; or of the
climate in question with that of another far distant.
This is called the *mean of the climate.*

Mean of the Climate of London.

If we regard the latitude, and elevation above the
sea, of London, independently of local circumstances,
the temperature has been hitherto rated too high; as

was that of the city itself in the earliest observations.
In the " Meteorological Journal kept at the Apartments of the Royal Society by order of the President
and Council," the period from 1778 to 1781 gives a
mean of 52° 65. In 1787, this register being resumed
after a cessation of five years, we have an account of
precautions now used to secure accuracy, and the ten
years from 1787 to 1796 make it . ... 50° 516 :
A similar period to 1806 inclusive ...... 50° 490 :
A third, ending with 1816 .... 50° 364.

Mean of the City on the average of thirty
   year's observations .... 50° 456.

But the temperature of the *city* is not to be considered as that of the *climate :* it partakes too much of
an artificial warmth, induced by its structure, by a
crowded population and the consumption of great
quantities of fuel : as will appear by what follows.

My own observations were conducted for the first
three years at *Plaistow ;* the site being about $3\frac{1}{2}$ miles
NNE of the Royal Observatory at Greenwich. The
village is four miles east of the edge of London : it
has the Thames a mile and a half to the S, and an
open level country, for the most part well drained
land, around it. The thermometer was attached to a
post set in the ground, under a Portugal laurel, and
from the lowness of this tree the whole instrument
was within three feet of the turf : it had the house
and offices, buildings of ordinary height, to the S and
SE, distant about twenty yards, but was in other
respects freely exposed.

The average of all the observations at this station
for 1807, 1808, 1809, is .... 48° 848
  The same for London (Phil. Trans.) .... 50° 608
        London warmer 1° 760.

For the next three years, the observations given in
my First Volume were made, partly at Plaistow, and
partly at the Laboratory situated at *Stratford,* a mile
and a half to the NW; on ground nearly of the same
elevation. Some of these, probably, have derived an
excess of warmth from the contiguity of the instrument to a large building, in which many fires were
kept : others are doubtless somewhat too low, in consequence of a change which I made in the position
of the instrument at Plaistow, and which I found to
have the effect of depressing the maximum. The
thermometer at Stratford had an open NW exposure,
at six feet from the ground, close to the river Lea.

The average of these observations for 1810, 1811,
and 1812, is .... 49° 480

That at London for the same period .... 50° 949

London warmer 1° 469.

Tottenham Green, where my latter observations
have been made, is four miles from the North side
of London, and the country to the NW especially
being somewhat hilly, and more wooded, I consider
it as more sheltered than the former site ; the elevation of the ground is a trifle greater, and the Thermometer was also placed higher, being about 10 feet
from the general level of the garden before it, with a
very good North exposure ; but it was not quite
enough detached from the house, having been fixed
to the outer door case, in a frame which gave it a
little projection, and admitted the air behind it. The
former instrument having been broken, this was a
different one, inclosed at first in a glass tube of an
inch and a half diameter, the front of which I soon
caused to be laid open, to procure a more free radiation of the heat from the instrument within.

The average in this situation for the years 1813, 1814, 1815, and 1816 is .... 48.°233
And for London during the same period 49.°741
London warmer 1.°508.

Thus under the varying circumstances of different sites, different instruments, and different positions of the latter, we find London always warmer than the country, the average excess of its temperature being 1° 579.   But as the same causes which produce an artificial elevation of temperature in London, must likewise influence in a smaller degree the country, the mean of which for the ten years ending with 1816 is 48,79, and as the second fractional figure was uniformly neglected in taking the monthly means for the annual average in the Register of the Royal Society, I shall for the present abate a little of the one, and add to the other ; and for the purposes of comparison rate the *mean of the Latitude and level* of London at 48° 5, and that of the *Metropolis itself* at 50° 5. Future observations with Thermometers previously compared and a greater degree of care to secure the fractions, may determine these with an accuracy not as yet attainable.

Mean of the Year, and its variations.

I shall have frequent occasion in the course of this volume to present the reader with a series of results expressed by a curve ; a mode of speaking to the eye which greatly facilitates the comparison of such variable quantities, when we wish to contemplate them only as becoming greater or less, and to view the order of their increase and decrease without reference to the exact amount of the sums compared.

Fig. 1.

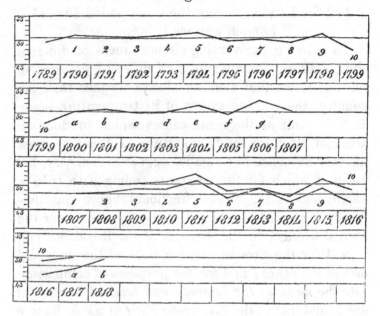

The flexuous lines in Fig. 1 are intended to shew in
this way, the variation of the Annual mean Temperature in the climate of London for the series of years
from 1789 to 1818 inclusive. The three upper curves
are deduced from the results of the register in the
Philosophical Transactions ; the lower one, extending
from 1807 to 1818, from the observations detailed in
my first volume, with the addition of two years published in Thomson's Annals of Philosophy, and since
inserted in the present Volume. The results having
been first marked over their respective years, on a
scale formed by lines ruled vertically for time, and
horizontally for the temperature, the curves were then
prolonged from one point to the other in succession.
The mean temperatures thus expressed will be found
in figures in the following Table.

Annual Mean Temperature.

		In London.	In the Country.
1789	...	49·491	
1790	..	50·892	
1791	...	50·833	
1792	...	50·483	
1793	..	50·820	
1794	....	51·200	
1795	...	49·700	
1796	..	50·083	
1797	...	49·398	
1798	..	50·999	
1799	...	47·920	
1800	...	50·522	
1801		51·080	
1802	....	50·200	
1803	..	50·329	
1804	...	51·731	
1805	....	49·998	
1806	....	52·734	
1807	...	50·733	.... 48·367
1808	...	50·466	.. 48·633
1809	..	50·633	.. 49·546
1810	...	50·976	.. 49·507
1811	..	52·666	... 51·190
1812	...	49·208	... 47·743
1813	...	49·741	. 49·762
1814	...	48·241	.. 46·967
1815	...	51·550	.... 49·630
1816	...	49·433	.. 46·572
1817	...	50·316	... 47·834
1818	...		.. 50·028

Averages.

		In London.	In the Country.
5 years from 1790	..	50·845	
5 years from 1807	....	51·095	.. 49·448
5 years from 1795	..	49·620	
5 years from 1812	....	49·634	.... 48·135
10 years from 1790	....	50·233	
10 years from 1807	...	50·364	. .. 48·791
17 years from 1790	..	50·530	
17 years from 1800	..	50·600	
7 years from 1800	..	50·912	

Extent of variation.

In London in 30 years .. 4·814	In the Country in 12 years 4·618
Highest mean in 1806	Highest mean in 1811
Lowest mean in 1799	Lowest mean in 1816

The Mean Temperature varies, as the reader will
have seen, in different years to the extent of 4 degrees
and 8-tenths of Fahrenheit : a quantity certainly not
considerable, when we compare by sensation the
warmth of one hour of the day with another; yet
capable, when added or abstracted for the whole year,
of producing a decided difference in the seasons. We
must not, however, too hastily connect with a low
mean the idea of a cold winter, or that of a warm
summer with a high one : the heat is added or taken
away sometimes in one season, sometimes in another,
and again occasionally almost throughout the year.
But it is worthy of notice, that notwithstanding the
great difference which we all find by sensation, in the
warmth of the same month, week, or day in different
years, summer and winter on occasion almost exchanging places, yet the total result of the seasons is
so nearly uniform in each, that no one year in thirty
is found by the most accurate mode of comparison to
differ from another quite five degrees, and the variation from year to year is usually not half as much.

To proceed from the amount to the *manner* of the
annual variation : it is for the most part such, that
the elevations and depressions take place in *alternate
years*, though some of them go on for two years;
and this tendency to alternation is still compatible
with a disposition to rise or fall on the whole through
a series of years. Thus in the space from 1794 to
1799, the mean is depressed three degrees, and from
1811 to 1816 (by my own observation,) four degrees
and a half: on the other hand, there is an intermitting
elevation carried on, from 1799 to 1806, by which in
the whole four degrees and a half are gained.

Lastly, and what is more important, there is evidence, which a few years more will perhaps render

conclusive, of the existence of *alternate periods of years* in the variation.

For the reader's help in comprehending this, I have numbered in the diagram the *ten* years of a period, which appears twice in the series here recorded; and have distinguished by letters the *seven* years of another; which having completed its course between the two periods of ten years, appears to have begun again immediately after the latter of these, and to be now in progress.

To begin with 1790, we have four years of an equable heat, upon or a little above the mean of London: the same equable average years will be found, in both the London and country observations, if we begin with 1807. Then occur six years alternating in temperature, from 1794 to 1799, the first of them the highest, the last the lowest of the ten to which they belong: the same circumstances obtain in the six years from 1811 to 1816. Or, if we compare the averages deduced from these two sets of ten years, as given in the Table, p. 95, we shall find 5 years from 1790 = 50.845, and 5 years from 1795 = 49.620; difference 1.225: again, 5 years from 1807 = 51.095, and 5 years from 1812 = 49.634; difference 1.461. That is, the latter half of each period is colder than the former by nearly the same quantity; while the two periods entire, average, within an inconsiderable fraction, alike. The period of seven years from 1800 to 1806, I have already noticed as an ascending series: in this, two of the elevations go on through two years. I consider it as having probably recommenced in 1817, because that year rises above 1816, and the following year, 1818, above both; as 1800 and 1801 do above 1799.

A chasm in the Register of the Royal Society im-

mediately previous to 1787, prevents me from bringing into the parallel a series of seven years antecedent
to 1799. If this series was on the whole an ascending
one, it scarcely could have ended with 1789, which is
stated at a degree below the average. On the other
hand it is certain from different Registers, that 1782
was as far below the average of the climate, as 1799
and 1816. The year 1787 is stated at 51 02, and
1788 at 50·63, in which depression of the mean they
agree, but not in due proportion, with the corresponding years *e* and *f*, in the middle series.

On the whole, the want of observations with a self-
registering Thermometer before the year 1794, throws
some degree of uncertainty upon those early results;
though it is probable none of them err a degree from
the truth, at least if we put the artificial warmth of
the city out of the question. Six's Thermometer,
after having been in use at Somerset house for seventeen years, was disused towards the close of 1810:
and perhaps I may not unreasonably attribute to this
cause the discrepant proportions of the London curve
for the following years, in one which the temperature
of the city loses its accustomed excess, and is even a
small fraction below that of the country.

Should the results of the present, and four following
years, to 1823 inclusive, correspond sufficiently with
c, d, e, f, g, the inquiry respecting these alternating
periods may be resumed, taking in all the evidence
that can be procured from early registers, and even
carrying it into the corresponding years in the
Meteorological Journals of other countries : for it is
clear that the causes of such periodical changes in a
climate must be astronomical, and not local : and this
circumstance, if established, must lead us to expect
occasional irregularities, and as it were intercalations,

in the periods, which a long series of years can alone satisfactorily explain.

Of the connexion of a high or a low mean Temperature with the state of the Barometer, the Rain, and other phenomena of the year, it is too early to treat at present: it is sufficient to have shewn from the manner and proportions of the variation of the annual mean, that this variation is probably periodical, or that annual mean temperatures nearly approaching to each other occur at intervals, consisting of definite periods of years.

Mean of the Month, and its variations.

From the variations of the mean heat during a series of years, we may proceed to its distribution among the several months of the year, and the variations of the mean for each of them.

The general Table A, at the end of the volume, exhibits the mean Temperature of each month, in each of twenty years, ten of which, from 1797 to 1806, were taken in London, and the remainder, from 1807 to 1816, in the country.

The *averages* of these mean temperatures come out as follows.

Mo.	For the city. 1797—1806.		For the country. 1807—1816.		For the whole. 1797—1816.
1 Jan.	38·52	....	34·16	....	36·34
2 Feb.	39·42	....	39·78	....	39·60
3 Mar.	42·51	....	41·51	....	42·01
4 Apr.	48·31	....	46·89	....	47·61
5 May	55·01	....	55·79	....	55·40
6 June	60·07	....	58·66	....	59·36
7 July	63·45	....	62·40	....	62·97
8 Aug.	64·45	....	61·35	....	62·90
9 Sept.	59·18	....	56·22	....	57·70
10 Oct.	51·33	....	50·24	....	50·79
11 Nov.	43·86	....	40·93	....	42·40
12 Dec.	39·76	....	37·66	....	38·71
difference of the extremes ...	25·93		28·24		26·63

The warmest month in the year therefore differs in its mean temperature from the coldest, *on a long average*, about twenty-six degrees and a half of Fahrenheit; and this difference is greater by nearly two degrees and a half, in the country, than it is in London.

In this long average the inequalities of temperature in the same month, which constitute the principal difference of our seasons, are in great measure extinguished, the extremes balancing each other. The series of mean Temperatures in the third column presents, therefore, a near approach to that regular gradation of heat, increasing and decreasing through the seasons, which a consideration of the primary astronomical causes of summer and winter, in temperate latitudes, would lead us to expect. To make this

Fig. 2.

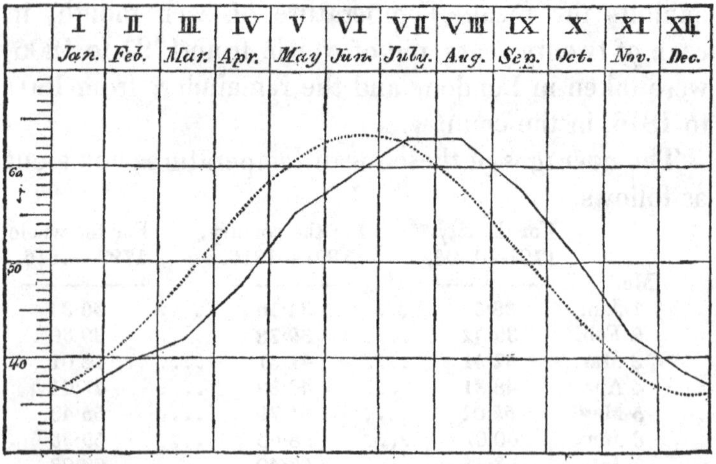

more obvious, I have placed in Fig. 2, a curve, constructed from the series of results in question, by the side of another represented by a dotted line, which latter expresses, on a scale of the same extent as that of the *temperature,* the progress of the sun in *de-clination* through the year. As I shall have occasion hereafter to enter more at large into this comparison, I shall only request the reader to notice, here, the manner in which the Monthly mean Temperature, following at some distance the elevation and depression of the sun, advances from its lowest point in winter, through the spring months, to its greater elevation in summer; and then returns by an opposite gradation, through the autumnal months, to the point from which it set out.

But if, taking up the general Table A, we look for the same regular gradation in particular years, we shall meet with many exceptions, attended still with some appearance of order and compensation. For instance, in 1797, the temperature of the Second month scarcely differs from that of the First, and both are below the average of that month: the Third has the average temperature of the Second: the Sixth is two degrees deficient, while the Seventh has two in excess: a deficient temperature then again prevails, till at the close of the year we have an excess of three degrees. And in the country observations, in 1807, after a correct average mean in the First month, we see the Second and Third as it were exchanging places, and both cold: the Seventh, on the contrary, has a warm mean, and the Eighth a hot one; which difference in the following year is reversed in those months: then (in 1807,) we have the Ninth and Tenth almost precisely equal, while in the following year, the latter month is the coldest by nearly nine degrees: lastly, 1807, goes out, as it came in, with an average mean. A careful perusal of the Table in this way, and still better the reducing of the several years to curves, on a scale similar to that on which I have placed the mean, will give the reader an adequate conception of the manner in which the comparative coldness of one month, or season, is balanced by the warmth of another, and *vice versa;* while some years are warm and others cold nearly throughout.

If the monthly means in this Table be examined for the same month in successive years, down the column, it will be perceived (consistently with what has been stated respecting the Annual mean), that together with alternations in temperature, there are occasional gradations carried through several years, towards a warmer or a colder mean; while in a few instances, the warmest and coldest months in the series lie almost together. The greatest *extent* of these variations is marked at the foot of the column; and it is observable, that while the year scarcely differs in its mean temperature *five* degrees at most, the month is subject to a variation which in several cases amounts to *ten,* and in one runs up to *fourteen* degrees. I very well recollect, and have verified, the extraordinary warmth of the month of December 1806, on which the latter result depends: my own observations at Plaistow make the mean of this month 45·30, which, with a full allowance for the winter excess of the city temperature, comes nearly to the same thing. This year was the highest of an ascending series of seven, which I have marked with letters in Fig. 1. It was warm nearly throughout, and the heat was most in excess at its close: the cause of the excess therefore was neither a local nor a transient one.

Mean of the Month in London and in the Country, with their variations compared.

I have already stated that London has an artificial excess of heat, and shewn the average amount of this excess in the whole year. In examining the monthly *means*, to see whether it was alike throughout the year or unequally distributed, I found the latter to be the case; and that attended with circumstances of considerable interest.

Average Monthly Mean Temperature 1807—1816.

Mo.	In the Country.		In London.		London warmer.
1 Jan.	34·16	..	36·20	....	2·04
2 Feb.	39·78	....	41·47	....	1·69
3 Mar.	41·51	....	42·77	....	1·26
4 Apr.	46·89	....	47·69	..	0·80
5 May	55·79	.. .	56·28	....	0·49
6 June	58·66	.. .	59·91	....	1·25
7 July	62·40	..	63·41	....	1·01
8 Aug.	61·35	..	62·61	.. .	1·26
9 Sept.	56·22	. .	58·45 *	:...	2·13
10 Oct.	50·24	....	52·23	....	1·99
11 Nov.	40·93	:.. .	43·08	..	2·15
12 Dec.	37·66	....	39·40	....	1·74

That the superior temperature of the bodies of men and animals is capable of elevating, in a small proportion, the mean heat of a city or populous tract of country in a temperate latitude, is a proposition which will scarcely be disputed. Whoever has passed his hand over the surface of a glass hive, whether in summer or

* In taking out the London results for this average, I was obliged to reject that for the month of September 1815; many of the observations in this month being manifestly erroneous, and the mean at least 6° too high. The average of the month for the first five years exceeds that of the country by 1·77 only.

winter, will have perceived, perhaps with surprise, how much the little bodies of the collected multitude of bees are capable of heating the place that contains them : hence in warm weather we see them ventilating the hive with their wings, and occasionally preferring while unemployed to lodge, like our citizens, about the entrance.

But the proportion of warmth which is induced in a city by the population, must be far less considerable than that which emanates from the fires : the greater part of which are kept up for the very purpose of preventing the sensation attending the escape of heat from our bodies. A temperature equal to that of spring is hence maintained, in the depth of winter, in the *included* part of the atmosphere, which as it escapes from the houses is continually renewed : another and more considerable portion of heated air is continually poured into the common mass from the chimnies ; to which, lastly, we have to add the heat diffused in all directions, from founderies, breweries, steam engines, and other manufacturing and culinary fires. The real matter of suprise when we contemplate so many sources of heat in a city is, that the effect on the thermometer is not more considerable.

To return to the proportions held by the excess of London, it is greater in winter than in summer, and it sinks gradually to its lowest amount as the temperature advances in the spring : all which is consistent with the supposition, that in winter it is principally due to the heat diffused by the fires. An addition of one or two degrees being of more value on a low temperature than on a high one, I reduced the numbers in the third column of the Table to the fractional parts which they give upon those in the first ; when they came out thus, beginning with the First month :

$$\tfrac{1}{17}, \tfrac{1}{24}, \tfrac{1}{33}, \tfrac{1}{59}, \tfrac{1}{114}, \tfrac{1}{48}, \tfrac{1}{62}, \tfrac{1}{49}, \tfrac{1}{26}, \tfrac{1}{26}, \tfrac{1}{20}, \tfrac{1}{22}.$$

We have here a near approach to a regular gradation, the *proportion* of excess on the lower temperature decreasing from the First to the Fifth month and then increasing again to the First. But the relations of the respective mean temperatures, with other circumstances attending them, will be best seen by means of curves.

Fig. 3.

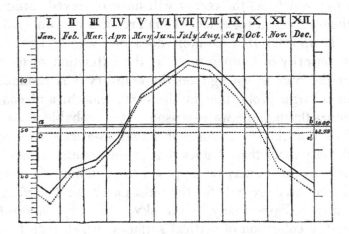

Fig. 4.

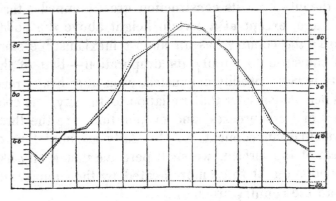

o

In fig. 3, the full line represents the monthly mean of London, as given in the Table, the dotted line that of the country : the horizontal lines *a b c d* are on the respective annual means ; and both curves are on the same scale.

In fig. 4, the respective curves are laid down on se-parate scales, and that for the country temperature, which is still a dotted line, *is elevated* 1° 57, or the amount of the mean annual difference between the two ; which, as the reader will have observed, brings them very near together.   I shall remark first on fig. 3.

It appears that London does not wholly lose its superiority of temperature, by the extinction of most of the fires in spring : on the contrary it is resumed in a large proportion in the Sixth month, and con-tinues through the warm season.   It is probable, there-fore, that the sun in summer actually warms the air of the city more than it does that of the country around. Several causes may be supposed to contribute to this : the country presents for the most part a plain surface, which radiates freely to the sky,—the city, in great part, a collection of vertical surfaces, which reflect on each other the heat they respectively acquire : the country is freely swept by the light winds of summer, —the city from its construction greatly impedes their passage, except at a certain height above the build-ings : the country has an almost inexhaustible store of moisture to supply its evaporation—that of the city is very speedily exhausted, even after heavy rain. When we consider that radiation to the sky, the con-tact of fresh breezes, and evaporation, are the three principal impediments to the daily accumulation of heat at the surface, we shall perceive that a city like London ought to be more heated by the summer sun than the country around it.

But this effect is not produced very suddenly. For while, in the forenoon, a proportion of the vertical walls are exposed to the sun, the remainder are in shade, and casting a shadow on the intervening ground. These are receiving however, in the wider streets, the reflected rays from the walls opposed to them; which they return to the former, when visited in their turn by the sun. Hence in the narrow streets, especially those that run E and W, it is generally cooler than in the large ones and in the squares. Hence too, in the morning of a hot day, it is sensibly cooler in London than in the country, and in the evening sensibly warmer. For the hottest time in a city, relatively to the hour of the day, must be that, when the second set of vertical surfaces having become heated by the western sun, the passenger is placed between two skreens, the one reflecting the heat it is receiving, the other radiating that which it has received. Many of my readers must recollect having felt the heat of a western wall, in passing under it long after sunset.

Let us now advert to the curves in fig. 4, in order to be convinced that the same cause operates also, on the great scale of the year. In this figure, by elevating the lower scale, we have done away the mean difference of 1° 57 in the annual temperature; or in other terms made the country as warm as London. It will now be seen that the remaining difference consists principally in this: that for six months of the year, from the Second to the Eighth inclusive, the country curve holds the higher place, and for the remaining six months, the London one. This proves that, although London is always warmer than the country, *the former acquires and loses its heat more slowly than the latter,* being left behind both in the

ascending and descending scale. To the same cause
we may probably ascribe the remarkable fact, which
appears on the average of twenty years (though not
in the series of *ten* of which we have just now treated)
that although the Seventh be the hottest month in the
country, and on the whole average, the Eighth month
exceeds it in temperature by just one degree, in London.

Extremes of the Climate.

Before proceeding to investigate the variations of
the diurnal temperature, which will conduct us through
the seasons, and complete this part of the subject, it
will be proper to devote a few pages to the *extremes*
of Temperature to which the year, month, and day
are respectively subject.

The General Table B, at the end of the volume,
exhibits the highest and lowest temperatures observed
monthly in the country (where alone these points can
be accurately ascertained) during the years from 1807
to 1816 inclusive. I have annexed to each observation the prevalent wind or winds at the time, and in
some cases, where it is considered to have equally influenced the temperature, the wind of the day preceding the observation. The maximum and minimum
of the year will be readily found by the marks * and †
affixed to them.

Of the extremes of cold, the far greater number occur in the First month, only two being in the Twelfth
and one in the Second. The extremes of heat are
more diffused : only five of them fall in the Seventh
month, and the remainder in diminishing proportion
earlier and later in the summer. Thus of the whole
twelve, there are only two months in spring and two
in autumn, which are not occasionally subject to one
or the other annual extreme of Temperature.

The Thermometer stood in the year

Year		High		Low		Range		Medium
1807	at	87°	and	13°	....	74	....	50
1808	....	96	....	12	....	84	....	54
1809	...	82	....	18	....	64	....	50
1810	....	85	....	10	....	75	.. .	47·5
1811	....	88	...	14	....	74	....	51
1812	....	78	.. .	18	....	60	....	48
1813	....	85	....	19	...	66	....	52
1814	....	91	....	8	....	83	..	49·5
1815	....	80	....	17	....	63	....	48·5
1816	...	81	....	—5	....	86	...	38
Averages	.....	85·3		12·4		72·9		48·85

I have before stated the mean Temperature of the country, on the average of all the *daily* extremes, or which amounts to the same thing, of the medium of each day, during the above ten years, at 48° 79. On the average of all the *monthly* extremes in Table **B** it is 48° 34 : and on that of all the *yearly* extremes as given above 48° 85. Even the greatest heat and greatest cold, in these ten years, diverge to nearly equal distances from the mean of the climate.

This agreement in the different averages is certainly remarkable : it gives further probability to the opinion that these years form a series, comprising a revolution of Temperature complete in itself: it is likewise a striking proof of the utility of a self-registering Thermometer. It is possible that in the Thermometer of Six we possess an instrument, which being merely fixed to a post, and properly defended from the sun's rays and from accidents, in an uninhabited country, where it could be visited and adjusted by navigators once in every year, would give in a moderate run of years, with considerable accuracy, the mean temperature of the latitude and elevation where it stood. In

like manner might an accurate comparison be made
with little labour, at the summit and at the foot of
mountains, of the mean temperature of the several
months at different elevations. When Meteorology
shall become a science and be studied by navigators,
travellers, and men of competent skill engaged in local
surveys, experiments of this kind will perhaps be as
common as the taking of levels and angles and observing the motions of the heavenly bodies, for the
perfecting of Geography and Astronomy.

To return to the extremes of our own climate for
the last ten years—the day of *greatest heat* within my
observation was the 13th of the Seventh month, 1808,
when I was attentive for many hours to the phenomena; of these the reader will find notes, (which

Fig. 5.

Seventh Month, July, 1808.

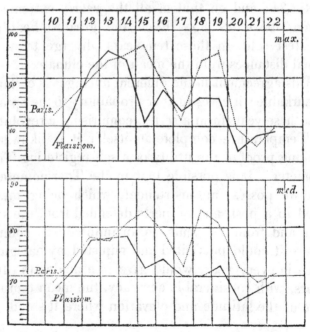

would have been more copious had I been aware at
that time of their importance) under Table 21 in the
first volume. To prove the extensive action of the
combined causes of this excessive heat, I shall here
compare by means of the curves in fig. 5, the maximum and medium temperature at Plaistow with that
at *Paris* (distant 180 miles to SSE) for the space of
thirteen days in which the principal elevation of temperature took place.

The maximum at Paris on the 10th of Seventh
month (see the Notes abovementioned) was 82° 6,
the wind NW : that at Plaistow on the same day was
76°, the wind SW. During the three following days,
the heat at each station increased steadily, the wind
at Paris being E & SE, and at London S & SW   On
the 13th, when the thermometer with us had risen to
96°, the evening atmosphere presented *dew* to the SE,
and some traces of *thunder-clouds* to the NW : the
change then was approaching from the latter point,
while the atmosphere of Paris remained as yet undis-
turbed : its heat was below ours, being only 93° 8, and
it did not reach its climax till the 15th, when the
thermometer there rose to 97° 2, and fell only to 70°
at night, the wind SE. In the mean time distant
thunderstorms to the Westward, and one in particular
about Gloucester, of a character for intensity suited
to the exaltation of the predisposing causes, reduced
our maximum in two days to 81°, with the wind at
NE. In two days more the change thus propagated
from the Westward appears to have reached Paris :
they were cooled down on the 17th, to 81° 5 by day
and 62° 7 by night. Immediately after this, a second
elevation of temperature took place with them, which
was felt also in a less degree at London, on the 18th
and 19th : lastly the heat at both places went down to

the ordinary summer standard, by a SW wind introducing rain.

The mean heat of this period of thirteen days at
Paris was, by day 87° 69, by night 63° 92, on the
whole 75° 80 : the mean at London (Plaistow) was
by day 84° 38, by night 58° 77, on the whole 71° 57.
Thus Paris had on the whole (consistently with its
more southern latitude) about $4\frac{1}{4}$ degrees more heat
than London, yet with variations throughout strikingly analogous to our own : till on the 22d, with the
same winds in play at each station, the temperatures
of the two become nearly equal : in which situation,
although the comparison might have been further prosecuted, we may leave them.

This heat was not with us of the sultry oppressive
kind which commonly ushers in a thunderstorm : the
sky was serene and a fine breeze prevailed, yet such
was the ardour of the sky, that motion was unpleasant
and labour in the sun dangerous : the feather'd tribes
were mute by day, and revived by the freshness of the
night were heard singing by moonlight. In the evenings the *dew* fell pretty freely, and at temperatures
which in ordinary circumstances would have sufficed
instantly to dissipate it : but the production of this
phenomenon depends at all times, not on the absolute
but on the relative temperature of the calm evening
air after a warm day ; and if this be cooled 20° or 25°,
it matters not whether it were previously charged with
water at 55° or 95°, provided the refrigeration pass
that degree at which the whole quantity can no longer
subsist as vapour.

I had equal opportunity of observing at Tottenham
the *intense cold* of the 9—10th of Second month 1816,
respecting which I need not enlarge here, having
given already a pretty long note on the subject, under

Table 115, vol 1. We had on this occasion likewise in the day time a clear atmosphere: a gale from the NE had precipitated in snow the moisture which previously abounded, and which had twice in the space of a few days brought the hygrometer to 100° So cold was the surface on the 9th at noon, that a bright sun, contrary to its usual effect in our climate, produced not the least moisture in the snow, the polished plates of which retaining their form, refracted the rays with all the brilliancy of dew drops: the thermometer in these circumstances reached only to 20°, or *seventy-six degrees* below the temperature of the middle of the hot day I have described : in the night it went down (in its usual position) to minus 5, and there is every reason to believe that the mass of our atmosphere was on this occasion at a temperature below zero for about twelve hours. This is a state of the air not uncommon, I believe, for several days together, on the continent in higher latitudes, but with us it is happily, of necessity, very rare and transient.

A comparison of the observations at Tottenham with those at Paris will again furnish some curious coincidences. I shall present the minimum and medium of each in curves, as before, and insert here the temperature at Paris reduced to Fahrenheits scale, from the observations of Bouvard in the Journal de Physique : my own as to the daily extremes, will be found in their place in the first volume : the daily *medium* I shall annex here.

The mean temperatures at the two places for the period occupied by these observations, appear to bear very nearly the same proportion to each other as in the former series; Paris being 4° 67 warmer than London.

Fig. 6.

Second Month, February, 1816.

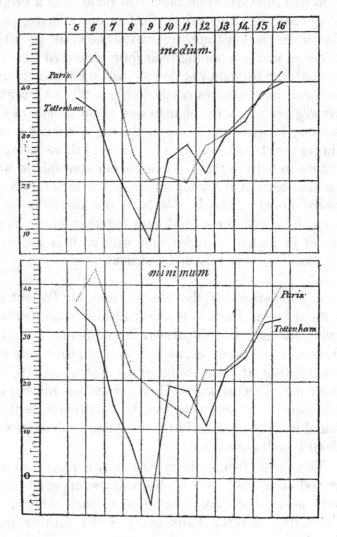

		At Paris.				At Tottenham.	
1816.	MAX.	MIN.	MED.	WIND		MED.	WIND.
Second Mo. 5	45·68	36·50	41·09	W	....	37·00	S
6	48·65	43·25	45 95	SW	....	34·50	SE
7	48·65	32 45	40·55	SW	....	23·00	NE
8	30·20	21·65	25·92	NE	....	15·50	N
9	22·10	18·25	20·17	NE	....	7·50	E
10	26·15	15·35	20·75	ENE	....	24·50	SW
11	27·05	12·65	19·85	E	....	27·50	N
12	35·60	22·55	29 07	NE	....	21·50	N
13	36·50	22·55	29·52	NE	....	29·00	Var.
14	41·00	26·15	33·57	N	....	32 00	W
15	41·45	33·65	37·55	WNW	....	38·00	SW
16	44·60	39·65	42·12	W	....	40·00	NE

Mean of 12 Days ..... **32·17**   27·50

A slight *elevation* of the nocturnal temperature on
the 5th at Tottenham, was followed by a corresponding but more considerable one, at Paris on the 6th :
it was misty with some rain at both. On the evening
of the 6th, the rain with us by the change of the wind
became sleet, and finally snow, which fell in the night
in great quantity, and at intervals in the day after ;
while there was still only rain at Paris. The North-
Easterly gale did not set in there until the 8th, and
the depression of temperature, with the snow, was late
in proportion. This depression, which on the first
days proceeded as rapidly with them as with us, appears to have experienced a check, at the time when
our own temperature was at the lowest, and we find
the *extreme* of cold at Paris *two* days later than at
London : it is moreover not by any means proportionate to the difference in the mean temperature of
the two places, reaching only to about 13°.

In the subsequent rise of the Thermometer we see
Paris take the lead of London, contrary to the order
in the beginning of the series : there is a second depression also in the Tottenham curve, from a cause
which is felt more slightly and somewhat later at

Paris. Lastly, when the frost is actually going off and a Westerly wind makes its appearance at both stations, we see the elevation of temperature at each go on together.

Probable conditions of each extreme of Temperature.

Let us now review the case of each extreme of temperature in our climate, and see what conditions appear necessary to its production : and first of the *extreme of heat.*

To produce the highest possible temperature in our climate, there appears to be required,

First, a *clear atmosphere* at the time : that the sun's rays may have the freest possible access to the earth's surface.

Secondly, a *dry* and *warm* state of the *soil*, to some considerable depth ; that the earth may reverberate freely, without throwing up such a quantity of vapour as by its speedy condensation, in the higher and colder regions of the atmosphere, might produce cloudiness and annul the first condition.

Thirdly, these two causes must concur at a season when the sun is not far from his greatest elevation : otherwise the heat will be in excess, only relatively to that season of the year in which it occurs.

Fourth and lastly, to carry the heat to the very highest point, we must receive, at this crisis, by means of steady southerly breezes, the air of the Southern parts of Europe, while these in their turn are supplied from Africa, and the South of Asia. A wind of this kind, which would travel from Paris to London in a day, would reach us in a week from the tropic of Cancer.

To produce the *lowest possible temperature* in our climate there is required,

First, as in the former case, a clear and dry atmosphere at the time, that the heat may freely escape by
radiation : this condition will be best appreciated by
those who have read the experiments of Dr. Wells, on
the subject of the radiation of heat from the earth's
surface.

Secondly, a cold state of the soil, (the usual result
of previous cloudy, wet and frosty weather), and this
to some considerable depth : that the sun's rays may
not be assisted by any warmth from beneath, in raising
the temperature by day.

Thirdly, the concurrence of these two causes with
a sufficiently low state of the sun, and consequent
length of night : otherwise the cold, although severe
for the season, will not be such as to be remarkable in
comparing together the results of a series of years.

Fourthly, a cause must concur, which but for the
parallel which I wished to exhibit, between certain
conditions common to the two cases, I should have
placed first—the winds must come to us from the
Northward ; when if they blow with sufficient steadi-
ness, we may receive them at length from Siberia.

When this state of the wind has supervened upon
our mild winter weather, it speedily gives us a serene
atmosphere : our moisture is first precipitated on the
meeting of the two currents, in an abundant snow, the
latent heat of the vapour being given out to the air,
which passes with it to leeward. The air which suc-
ceeds, coming from a still colder region, and being
intensely dry, our own ice and snow evaporate into it,
and there is thus, towards the close, a contrary effect—
an absorption of heat, which undoubtedly contributes
to carry the depression to its extreme point.

It may seem extraordinary, in the case I have just
reviewed, that at such a distance from the winter

solstice, the power of the sun should not have prevented, in a greater degree, the effect of the Northerly
current. But we have probably, here, an effect similar
to that which takes place in a single night of frost:
the temperature (as is well known) is then often lowest just before sun-rise, the nocturnal depression, an
effect of the sun's absence, continuing to go on until
the approach of his rays becomes again sensible: so
in the present case, the long time during which the
sun had been low, may be admitted among the predisposing causes of the extreme depression—which in
all probability, would not have been produced by the
concurrence of the other causes, at an equal interval
before the solstice. I find that the depression in 1796
to — 6,5 which I mentioned in the note to Tab. 115,
as analogous to the present, occurred in the night
between the 24th and 25th of December*, just after
the winter solstice: it was preceded by a heavy snow
on the 23d, and a clear day on the 24th, with the temperature at 23° at noon.

Such are the causes, the concurrence of which appears requisite for the production of the extremes of
temperature in our climate: and they will probably
be found to obtain, more or less, in most of the cases
of great heat and excessive cold recorded in our
Registers. The history of the means by which the
equilibrium is restored, and the temperature made to
approximate again to the ordinary state of the season,
is more simple. It appears to be effected, *in both
cases*, by an irruption on land of the more temperate
air of the Atlantic. In effect, a SW wind was no
sooner established, than the heat in one case, and the
frost in the other, gave place to its influence.

---

* Extracts from a Meteorological Journal kept at Edmonton, Middlesex. By John Adams, 1814.

By admitting this reaction of the Atlantic atmosphere, we are able to solve the problem of the maximum of temperature, in one case, and the minimum in the other, falling two days *later* at Paris than at London ; in consequence of which, on a given day of the cold season, it was colder by several degrees at Paris than at London, and on another in the hot season, warmer at London than at Paris : for the latter city, lying more remote from the Atlantic than London, and in the midst of a larger mass of atmosphere resting on a continent, and more difficult to displace than our insular air, was in consequence later in receiving the change : and the causes, whether of elevation or depression of temperature, continued to operate during the interval.

To conclude, it will appear on examining the Table B throughout, that our warm weather *in winter* has almost uniformly come from the SW, the S, and W : but *in spring and summer* as frequently from the S and E : and that with regard to the cold extremes, a large proportion of them are connected with a NW wind, which in some instances is set down NW a N, (North-west *after* North). The reason of this connexion may be, that after the wind has been for some time North East or North, it has shifted to the NW just before the change to the Southward, and when (from a cause before explained) the cold by continuance arrives at its greatest intensity for the time.

Extremes of Day and Night.

The difference between the temperatures of day and night, or between the higher and lower extreme of the twenty-four hours, is subject to great variation. Sometimes from the effect of a steady wind with cloudy

weather, or of slight frosts with snow, the temperature
will scarcely vary five degrees in twenty-four hours:
at others, a clear night succeeding to a day of much
sunshine, or the sudden going off of severe frost, by
a change of wind, shall cause a variation in either
direction, of twenty, thirty, or more degrees. The
reader will find many examples of these changes, in
examining the periods about the middle of winter
and beginning of summer. To give a few instances:
Tab. 88, Dec. 4—9; six days with a uniform maximum of 44°, and the nights mostly but three or four
degrees colder: Tab. 40 and 41, many days in First
month very uniform: Tab. 21, three days at 92°, 96°,
94°, and the nights at 63°, 60°, 63°, respectively:
Tab. 144 (in the present volume) several nights, in
the fore part of the Sixth month, in which the thermometer was lower, from 35° to 37°, than in the day:
Tab. 116, Second month, 9—10, a rise from —5 to
30°: Tab, 136 (in this volume) Tenth month, 29—30,
a rise from 27° to 57°: again, Tab. 131, Fifth month,
15—17, the nights at 33°, the days at 65° and 67°.

Extremes of Day and Night in London and in the Country.

But it is by mean results in this, as in several previous cases, that we arrive at the clearest view of the
subject. In the two Tables, p. 134 to 137, I have
given under the titles *Higher mean*, and *Lower mean*,
the Monthly averages of the daily *maxima* and *minima*
of the Thermometer for twenty years: ten of them in
London, and the following ten in the Country. A
mean of these numbers being taken for each month,
on each set separately, the following results are afforded.

The higher mean, or *heat of the day,* taken on the observations from 1797 to 1806, *in London,* is 56° 17 : the lower mean, or *cold of the night,* on the same, 44° 80.

The higher mean *in the Country,* taken on the observations from 1807 to 1816, is 56° 51 ; the lower, 41° 10.

The mean variation of temperature, from the heat of the day to the cold of the night, is therefore

In London .... 11° 37
In the Country .... 15° 40
———
Greater mean variation in the Country 4° 04
———

The close coincidence in the averages of the *heat by day,* in London and in the Country, for two different decades of years, is certainly remarkable. I am prevented from forming an accurate comparison *on the same set* of years in either case, by the want of a complete series of observations with Six's Thermometer ; without which it is useless to attempt a parallel of the *extremes* of temperature.

Now, as to the *nights,* 44.80 — 41.10 = 3.70 : and we found before an excess for London, *on the mean of twenty-four hours,* of 1.47, which doubled (as it was halved by taking the medium) is 2.94. This difference in the average of the nocturnal extreme, exceeding the difference on the total average of mean temperatures, makes it probable that the excess of London, however acquired, is retained in such a way as to operate chiefly on the lower observation, entering but in a small proportion into the maximum. And, in effect, the averages at the bottom of the columns in the *Lower mean* table, shew that the nights in London, are at all times so much warmer than in the country,

Q

that no difference of seasons in ten years is able to reduce them below the latter in the average. Whereas, in the *Higher mean,* from the cause just mentioned, the monthly average of London is sometimes above, sometimes below, at others nearly parallel with the country one.

Extremes of Day and Night in the several Months of the year.

It is natural to expect that the difference between the temperatures of day and night should increase, in proportion as the sun acquires more power by elevation ; or that it should be greater in summer than in winter. The following Table, drawn from the two general ones to which I have just now referred, will shew to what extent, and in what proportions, this difference proceeds through the seasons. In constructing it, I have preferred the long average, which includes ten years in London : and have obviated the effect of the local warmth of the city, by deducting its excess, not from the mean of twenty-four hours, but, (on the ground of the preceding examination) from the *lower* mean exclusively. Thus, the average of the lower mean on ten years in London being for the first month 35·44; that of the higher mean 41·61; and the two for the country, respectively, 29·33 and 38·96, the calculation for this month runs as follows :

$$35·44 \quad - \quad 2·04 \text{ (see page 103)} \quad = 33·40$$
$$33·40 \quad + \quad 29·33 \quad = 62·73 \div 2 = 31·365$$
$$41·61 \quad + \quad 38·96 \quad = 80·57 \div 2 = 40·285$$
$$40·285 \quad - \quad 31·365 \quad = \quad 8·920,$$ the difference between the higher and lower mean of the month on 20 years : and so for the remaining months.

Mo.	Mean of greatest heat by Day.		Mean of greatest cold by Night.		Difference,
1 Jan.	40·285	....	31·365	....	8·920
2 Feb.	44·635	....	33·700	....	10·935
3 Mar.	48·085	....	35·315	....	12·770
4 April	55·375	....	39·420	...	15·955
5 May	64·065	....	46·540	....	17·525
6 June	68·360	....	49·750	....	18·610
7 July	71·500	....	53·840	...	17·660
8 Aug.	71·235	....	53·940	....	17·295
9 Sept.	65·665	....	48·675	....	16·990
10 Oct.	57·060	....	43·515	...	13·545
11 Nov.	47·225	....	36·495	...	10·730
12 Dec.	42·660	....	33·900	....	8·760
Averages	56·345		42·204		14·141

The third column presents, we must remember, a
series of differences between the average extreme
temperatures of day and night, divested as far as possible by compensation, of the disturbing effects of
different winds, of cloudy or clear days, of the presence or absence of rain or snow, and of the variable
pressure of the atmosphere.

We find accordingly in these numbers a gradation
which agrees pretty well with that of the Sun's declination. The greatest difference is found in that
month in which the Sun is highest, and longest above
the horizon ; the least, in that in which he is least
elevated, and makes the shortest stay with us. But
there are other circumstances, not so obvious, connected with the proportions of these numbers, which
we are not yet prepared to discuss. It may suffice
therefore to consider them for the present, as *an ap-
proximation* to a series, representing, in degrees of
Fahrenheit's thermometer, the mean quantity of heat,
actually produced by the direct and reflex action of
the Sun's rays, in each month of the year.

Diurnal Mean: variation of the daily heat through the Seasons.

Perceiving, very soon after I had begun to investigate the Temperature, the necessity of a fixed standard, with which to compare the very considerable variations in the mean heat of the same day, in different years, I determined on constructing a set of Tables applicable to this purpose. I then possessed Observations on the Thermometer in the country for ten years : and as it was certain that the temperature of the year did not reach both extremes of its variation in so short a period, it appeared needful to take into the average the ten preceding years from the Register of the Royal Society. The method employed in forming these tables was, to set down the higher and lower observation of each day, under the day, through the month ; then, to repeat the operation for the same month in the next year ; and so on for ten years. The average of the sums in each *column* then gave the mean heat of the *day* for ten years, and that of each *line* in the table, the higher and lower mean alternately for the *month*. The *monthly means for ten years* being then deduced, both from the final column and from the line of averages át bottom, the agreement of these, within certain limits, was considered as proving sufficiently the correctness of the calculation. A similar set of averages being likewise deduced from the country observations, the medium betwixt the two was taken for a general standard of the diurnal and monthly temperature of the climate. See the Tables, p. 138 and 143.

The circle of Temperature for the year being thus obtained in figures, I became curious to see how it would appear in a diagram, and what relation it

would bear to a circle, placed within a scale of the
same extent, and representing the sun's progress in
declination through the year. This inquiry, not to
trouble the reader with an account of its progress,
terminated in the construction of the scheme of temperature and declination, contained in the two copies
of Plate 1, facing the Title-page, which I shall now
proceed to describe. The reader will first avail himself of the copy least coloured ; as we have to do at
present only with the lines and figures.

The diagram, as the reader will have perceived at
once, presents a circular scale for the year, divided
(except where the termination of the months required
a difference,) into intervals of five days. Each of the
lines forming these divisions forms likewise a scale of
temperature ; being cut at equal intervals by the con-
centric circles, which are distant from each other five
degrees of Fahrenheit ; the *highest* part of the scale
being *within*, or towards the centre, the *lowest, without*, or towards the circumference. Just without the
circle representing 50°, is another formed by a strong
dotted line : this is placed on the *mean of the climate ;*
and with reference to the declination, it represents also
the *Equinoctial* or *Equator.* The circle representing
the sun's delination through the year, would be readily
found by its being so greatly eccentric : it is however
further marked by the word *Ecliptic*, and by the signs
of the Zodiac, with degrees of declination marked at
intervals. The North declination is made to proceed
towards the inner or upper part of the scale of Temperature, the South towards the outer part, or bottom ;
the extreme distance from the Equinoctial each way,
(or 23° 28′,) being equal to 15° of the Thermometer.

Having thus far explained the figure, I must now
request the reader's attention to the circle formed by

a flexuous line, which traverses the scale through the year, and presents the same eccentric appearance as the circle of declination. This is the curve of the daily mean Temperature, prolonged through points marked for each day on the scale. On a general view, the reader will perceive that, like the circle of declination, it is highest in Summer and lowest in Winter, and that it crosses the mean line twice, in Spring and again in Autumn : but not at the same time with the declination, being about a month later. If we trace the correspondence of the two circles, we shall find this difference in time to obtain throughout the year.

At the Autumnal Equinox, on the 23d of September, (to omit for a while the numerical designations of the months,) the Sun being in Libra, we have the diurnal Mean Temperature at 55°, or six degrees above the Mean of the year; to which it does not attain in its gradual descent, until the 22—23 October, when the Sun, advancing in South declination, has nearly reached the first degree of Scorpio.

Proceeding through the next two signs to the Winter Solstice, the declination, keeping in advance of the Temperature, arrives at its South extreme on the 22d of December, the Sun in Capricorn : but the Temperature does not reach its lowest point (at 34,45) until the 12th of the following month. And as the declination varies but little for a considerable space about the Winter Solstice, so we have here a sameness in the line of Temperature, which after a small elevation, almost revisits the cold extreme on the 25th January. The lowest Temperature of the year may therefore be said to occur, about the time when the Sun enters Aquarius. During this time, the declination having proceeded Northward a few degrees, the two circles coincide, and for a considerable space,

the variable curve of the Temperature intersects, at
intervals, the regular one of declination. As the season proceeds, the latter takes the lead in rising, the
divergence of the two lines increasing up to the Vernal
Equinox, 21st March, the Sun in Aries; thirty days
after which, at his entrance into Taurus, the Temperature is about to touch the mean of the year, which
it crosses on the 24—26th April. The ascending
Temperature now follows the declination, keeping
the same distance as before, to the Summer solstice,
(22d June, the Sun in Cancer,) but the Temperature
at this time is at 58° 85, or six degrees short of its
higher extreme. When the Sun, having passed his
greatest elevation, has declined a little towards the
South, the two circles coincide as before, and with
the same solstitial character (if I may so use the term,)
in the curve of Temperature; which continues here
for a longer time about the same level; insomuch that
the local excess of London causes the hottest *days* to
appear in the beginning of August; whereas in the
country they are the 12th and 25th of July; which,
with greater consistency places the hottest season
in the space between those days, or about the Sun's
entrance into Leo.

From this time the declination falls in the scale,
keeping in advance of the temperature, and the divergence of the two from each other increasing (as
before in Spring,) down to the Autumnal Equinox:
from which point, it will be recollected, we set out in
the comparison.

Thus, the average of each day upon the observations of twenty years, though made under the disadvantage of a local cause, irregularly raising the
Temperature in one half of them, has furnished a
practical proof of that which was before admitted in

theory, that *the diurnal Temperature,* abating the influence of temporary causes of variation, *is determined by the Sun's altitude at noon throughout the year.*

The curve of the Mean Temperature, we may observe, scarcely ever rises or falls uniformly for a week together; but is continually interrupted by deviations. Yet the general effect so nearly agrees with the progress of the Sun, that, were the circle of declination shifted, and its centre made to coincide with that of the curve of temperature, the latter would cross the former in more than fifty places, besides a great many in which they would be in contact. Setting aside the effect of the local excess of London, these deviations appear to be the result of the different winds, which prevail at the same season, in different years, producing very considerable elevations and depressions of temperature, which however do not perfectly balance each other in the average of twenty years chosen for my tables. I suppose that a very long average, or one taken from a real natural period of years, and in which local influence on the Thermometer should be avoided or allowed for, would bring out a curve much more nearly resembling the circle of declination. And it is now proper to observe that the latter is not a true circle. For, there being seven days more on the Summer than on the Winter side, a true circle would not have intersected the Equator at the Equinoxes, while it departed to an equal distance from it at either Solstice. It is therefore somewhat oblate, or partaking of the form of a circle of larger diameter, in the Summer months : and there is every reason to conclude, that the true theoretical curve of the diurnal Mean Temperature, will in the end be found to have the same disproportion between the half of the circle above, and that below, the Mean Temperature of the year.

Natural commencement and duration of the four Seasons.

The fact of the mean and extreme temperatures occurring with so regular a relation to the Equinoxes and Solstices, yet at so considerable a distance after them, has suggested to me a new and more natural demarcation of the limits of the seasons of our climate : which I have now, with the help of the *coloured* plate, to lay before the reader.

It is clear that in these latitudes we have four seasons, distinguishable by the rest and progress of nature in the vegetable world. We have a germinating leafing *Spring*, a flowering *Summer*, a fruit-bearing *Autumn*, a dormant naked *Winter*. Now, the difference of these from each other depending chiefly on the temperature, as to its elevation and the direction in which it is proceeding, in those parts of the year in which they severally take place, if we can divide the yearly circle of temperature in such a way as shall at once make its four parts symmetrical, and bring them more nearly to accord in time with the natural appearances abovementioned, a departure from the customary divisions of the " Quarters " of the year in our calendars, will by the natural philosopher at least be cheerfully tolerated.

Let us then remove the beginning of the seasons *fifteen days* in each case from their respective present situations, placing them at that distance *before* the Equinoxes and Solstices.

Spring will then begin the 6th of the Third month, *March*, at the temperature of 39° 94* (see the Table,

---

* The initial and terminal temperature of the season is taken, in every case, at a *medium* between the day on which the one season ends, and that on which the other begins : thus $39.67 + 40.22 \div 2 = 39.945$: and so of the rest.

R

p. 142—143), it will occupy 93 days, and will end on
the 6th of the Sixth month, *June,* at 58° 08—the temperature having risen 18° 14 degrees.

Summer will begin on the 7th of June, and it will
last 93 days; during which space the mean temperature will have risen from 58° 08 to 64° 75, or 6° 67;
and have declined again 6° 59 : it will end on the 7th
of the Ninth month, *September.*

Autumn, beginning on the 8th of September, at
58° 16 ; will have 90 days: during which, the mean
temperature will have declined 18.35 degrees, and it
will close on the 6th of the Twelfth month, *December,* at 39° 81.

Winter, comprehending 89 days (or in leap years
90) will begin December the 7th. During this season
the mean diurnal temperature having fallen 5° 36 to
34° 45, will have risen again 5° 49 or to 39° 94, on
the 5th of the Third month, *March,* the concluding
day of the season.

To make the symmetry and proportions of the seasons, as thus distributed, more obvious to the sense,
the plate has been coloured thus :—the space between
the line of the annual mean and that part of the variable curve of daily temperature which lies *above* it in
the scale, is made *red :* this space may be considered
as representing the *heat* of the year. The space betwixt the mean line, and the curve of the daily temperature lying *below* it, is coloured *blue,* and may be
considered as representing the *cold* of the year. The
remainder of the ground of the scale being filled up
with four colours, appropriate to the seasons, they are
thus marked out from each other like the countries in
a map. The Summer is seen at once to contain the
largest portion of the sensible heat of the year; which
after increasing to the middle of that season, decreases

again to the beginning of Autumn. In this season, the *heat* gradually goes out, and is succeeded in the middle by sensible cold, which becomes considerable by the end. Winter exhibits as large a proportion of the cold as summer did of the heat, and with the like increase and decrease. In Spring, we see the cold gradually go off, to be replaced in the middle of the season by warmth ; their respective proportions being like those which obtain in Autumn, while their positions are reversed. Lastly, by the beginning of summer (with which we set out) we see the heat increased to a degree sufficient again to constitute that season.

Cause of the difference between the Astronomical and the real sum of heat; or between the Sun's declination and the Temperature.

It remains for me to shew why the Temperature, both in its increase and decrease, is always a month behind the Sun.

The heat existing from day to day in the portion of our atmosphere next the Earth, is at no time the simple product of the direct action of the sun's rays on that portion. It has been found by experiments carefully conducted, and continued for a great length of time, that the direct action of the sun's rays, in a calm air, will raise the Thermometer an equal number of degrees, whether the time be the summer or the winter solstice, whether the temperature be at summer heat or near the freezing point.* It is therefore probable that the mass of the air is similarly affected, and that the proportion of heat which it derives

* See a copious paper on the subject by Flaugergues. *Journal de Physique,* Octobre, 1818.

from the direct passage of the rays is alike in all seasons. The accumulation of heat near the surface of the Earth, which we always experience from continued sunshine, is evidently due to the stopping of the rays at that surface; to their multiplied reflections and refractions, in consequence of which they are as it were absorbed and fixed, for a time, in the soil and in the incumbent atmosphere. By this process the Earth, when in a cold state at the end of Winter, *becomes gradually heated to a certain depth* as the warm season advances. On the other hand, when the Sun declines, in Autumn, the soil thus heated acts as a warm body on the atmosphere, and *gives out again the heat it has received.*

The Thermometer is therefore placed betwixt the Sun and a reflector, the Earth; and the heat which it indicates is at all times the product of the compound action of the two bodies. Now, if I place a flat skreen suddenly before a clear fire, I shall not need a Thermometer to learn, that at the first moment the skreen reflects no heat into the space between them: it requires first to be heated itself, that is to say, the rays which first fall on it are for the most part absorbed: but as soon as heated it reflects copiously. It is thus with the Earth's surface: it is a skreen behind the Thermometer, which absorbs heat during the Spring, and gives it out again in Autumn.

Were it not for this effect on the part of the Earth, the heat indicated by the Thermometer, would probably on a long average (to obviate the remaining irregularities, caused by clouds, rain, wind, and evaporation) be precisely at its maximum and minimum at the Solstices, and at the mean at the Equinoxes. For the power of the Sun is proportionate to the quantity of parallel rays falling on a given area of the

Earth's surface. And this quantity is greatest when they are vertical, and diminishes as they become more oblique ; till in a perfectly horizontal position of the rays, it is null. On this principle depends the superiority in heat, of noon over morning or evening, of our summer over our winter, and of the tropical over the polar regions. As the Sun advances in North declination, therefore, the heat we derive from him increases, actually in proportion to his altitude, but not sensibly ; because a part of it is required to heat the Earth, and is lost there by absorption. As he declines Southward in the Autumn, the heat we receive actually grows less in proportion, but not sensibly ; because we now receive back a certain quantity from the warm Earth. And it would appear that, were the Earth's surface at a mean temperature, and were the Sun's rays suddenly and totally intercepted for the time, it would require about thirty days to be cooled down 7 degrees, or the difference between the temperature by the Sun and that by the Thermometer, and about the same time to be heated to the former temperature, on their return.

To make this effect also more sensible, I have coloured, in one of the Plates, the spaces between the curves of declination and temperature, *blue* on the side of the year towards Spring, and *red* on that towards Autumn : the one to represent the cold produced by absorption in the former season, the other the heat derived from radiation in the latter.

HIGHER MEAN IN LONDON.

Year	1. Jan.	2. Feb.	3. Mar.	4. Apr.	5. May	6. June
1797	40·00	42·46	45·70	54·06	61·96	65·93
1798	43·16	44·78	48·45	59·10	64·32	72·80
1799	38·12	42·60	43·70	49·06	59·35	65·66
1800	41·64	39·28	44·12	56·36	64·31	64·83
1801	44·97	43·96	51·35	55·46	63·35	68·90
1802	38·09	44·67	49·61	58·63	60·54	67·56
1803	37·45	42·00	49·87	57·30	59·35	65·50
1804	47·84	43·00	48·09	51·73	66·77	71·80
1805	38·87	44·82	49·67	54·56	59·90	65·20
1806	46·00	47·53	46·54	51·10	65·67	70·90
Average of 10 years.	41·61	43·51	47·71	54·73	62·55	67·90

Year	7. July	8. Aug.	9. Sept.	10. Oct.	11. Nov.	12. Dec.
1797	74·80	70·03	63·55	53·61	48·03	46·32
1798	71·64	73·25	65·26	57·32	45·40	38·00
1799	69·06	67·61	62·70	54·35	48·26	36·32
1800	74·42	75·19	66·03	55·25	48·13	43·00
1801	70·93	73·38	66·63	57·74	45·46	41·06
1802	66·96	76·35	68·46	58·67	45·66	42·77
1803	74·67	72·64	63·43	56·19	47·70	45·38
1804	69·71	70·03	68·50	58·54	48·80	39·83
1805	69·09	71·96	68·56	54·19	45·56	44·00
1806	70·70	71·87	66·03	58·03	53·06	51·06
Average of 10 years.	71·19	72·23	65·91	56·38	47·60	42·77

the Year .... 56·17

HIGHER MEAN IN THE COUNTRY.

Year	1. Jan.	2. Feb.	3. Mar.	4. Apr.	5. May	6. June
1807	40·42	45·71	43·35	55·80	65·06	69·33
1808	41·12	41 55	43·00	50:50	70·29	68·43
1809	40·51	50·71	50·64	50·20	67·51	68·20
1810	38·58	44·64	49·51	57·13	61·06	71·96
1811	37·45	47·50	54·70	61·13	70·54	71·83
1812	41 54	47 58	46·93	52·20	62·87	64·20
1813	39·16	49·57	51·67	58·10	65·64	69·53
1814	37·19	39·10	43·74	61·63	61·03	64·93
1815	32·55	50·64	55·09	58·30	69·90	72·40
1816	41·16	40·69	46·00	55·20	60·90	67·36
Average of 10 years.	38·96	45·76	48·46	56·02	65·48	68·82

Year	7. July	8. Aug.	9. Sept.	10. Oct.	11. Nov.	12. Dec.
1807	75·19	74·67	62·13	60·03	42·83	41·25
1808	77·87	72·51	64·06	53·96	50·00	39·35
1809	69·87	69·22	64·23	57·12	44·96	45·42
1810	70·35	71·06	68·03	58·80	49·63	44·32
1811	69·80	67·80	68·00	63·00	51 50	44.48
1812	66·52	64·38	64·70	56·93	46·76	39·22
1813	74·22	72·06	66·23	55·96	47·70	42·19
1814	75·19	71·87	66·10	56·06	46·20	45·38
1815	71.48	72·16	67·66	58·00	44·76	42·03
1816	67·61	66·71	63·10	57·61	44·13	41·83
Average of 10 years.	71·81	70·24	65·42	57·74	46·85	42·55

the Year .... 56·51

Of the Temperature.

LOWER MEAN IN LONDON.

Year.	1. Jan.	2. Feb.	3. Mar.	4. Apr.	5. May.	6. June.
1797	34·64	32·21	34·00	40·76	45·96	49·20
1798	36·09	35·10	37·48	44·10	48·70	55·20
1799	32·06	33·82	34·96	39·06	45·48	50·43
1800	35·71	32·71	34·70	45·63	49·74	51·13
1801	37·13	36·82	40·80	39·83	47·25	52·80
1802	31·16	37·00	36·70	43·33	43·76	51·60
1803	33·09	34·57	38·90	43·53	46·67	52·60
1804	42·13	34·89	38·38	40·86	52·41	55·13
1805	33·48	36·53	38·35	41·40	44·96	50·20
1806	38·90	39·35	38·93	40·30	49·87	54·20
Average of 10 years.	35·44	35·30	37·32	41·88	47·48	52·25

Year	7. July.	8. Aug.	9. Sept.	10. Oct.	11. Nov.	12. Dec.
1797	56:16	53·58	50·36	44·29	38·76	39·06
1798	56·09	58·00	52·53	47·03	37·83	32·38
1799	55·58	53·38	50·20	45·00	41·10	32·29
1800	56·74	57·64	54·13	44·83	40·00	37·06
1801	55·09	57·35	55·60	47·70	38·46	33·93
1892	51·32	58·77	52·00	46·29	39·10	35·83
1803	57·90	56·51	46·86	45·96	39·70	40·19
1804	55·90	56·35	55·00	48·38	43·06	34·45
1805	55·09	58·03	54·86	45·00	37·96	37·51
1806	57·13	57·16	52·96	48·35	45·20	44·80
Average of 10 years.	55·70	56·67	52·45	46·28	40·11	36·75

the Year .... 44·80

LOWER MEAN IN THE COUNTRY.

Year	1. Jan.	2. Feb.	3. Mar.	4. Apr.	5. May	6. June
1807	27·87	31·03	28·93	36·20	48·51	48·50
1808	30·87	30·27	31 38	35·60	49·54	49·73
1809	32·32	39·14	36·64	36·23	46·51	49·30
1810	31·54	34·21	36·87	39·06	40·90	48·46
1811	27·83	36·67	37·29	42·26	51·67	51·33
1812	32·22	37·17	34·58	35·50	46·64	47·36
1813	30·32	37·78	36·25	38·63	47·80	47·76
1814	28·35	27·25	31·90	40·06	40·09	47·06
1815	20·87	38·32	39·35	38·83	47·54	47·83
1816	31·10	26·10	32·48	35·23	41·71	47·73
Average of 10 years.	29·33	33·79	34·57	37·76	46·09	48·50

Year	7. July	8. Aug.	9. Sept.	10. Oct.	11. Nov.	12. Dec.
1807	54·25	55·90	44·03	46·09	32·26	31·54
1808	56·51	54·51	48·76	40·58	38·26	30·58
1809	52·41	53·77	50·70	43·83	34·30	35·41
1810	52·16	52·19	50·10	43·22	39·06	35·38
1811	53·87	50·87	47·66	49·09	39·30	33·03
1812	51·06	51·29	46·20	41·90	36·30	31·80
1813	52·78	50·61	49·16	41·38	34·96	34·67
1814	54·32	52·48	45·26	37·67	33·50	35·03
1815	50·70	51·83	43·10	41·41	31·93	30·48
1816	51·87	51·29	45·33	42·29	30·40	29·96
Average of 10 years.	52·99	52·47	47·03	42·74	35·03	32·79

the Year .... 41·10

MEAN TEMPERATURE of every day in the Year

	First Mo. Jan.	Sec. Mo. Feb.	ThirdMo. Mar.	Four.Mo. April	Fifth Mo. May	Sixth Mo. June
1	37·15	39·65	42·05	44·35	51·00	56·00
2	37·90	40·45	43·75	45·50	52·25	56·70
3	37·70	40·85	43·50	45·40	51·30	59·05
4	38·20	38·17	42·20	46·85	52·55	59·55
5	39·30	38·05	40·37	46·80	53·40	59·45
6	38·20	36·70	40·00	46·87	55·10	59·30
7	37·00	35·40	38·87	47·95	55·30	59·95
8	37·20	35·30	39·30	48·80	54·90	59·90
9	36·30	39·00	39·15	48·05	·54·40	61·15
10	35·55	39·00	39·70	48·75	53·50	61·15
11	34·85	37·65	40·70	47·40	53 47	58·70
12	34·00	35·75	41·20	46·12	52·90	58·25
13	36·25	36·65	40·50	45·90	51·60	59·70
14	39·30	35·90	42·75	47·27	51·05	60·35
15	39·00	38·30	42·97	49·52	52·20	59·60
16	38·10	37·55	41·72	49·75	53·45	60·50
17	39·00	38·00	42·60	49·80	54·55	61·10
18	40·20	38·25	42·75	50·15	55·00	61·35
19	41·70	38·05	41·05	50·50	55·50	60·55
20	43·10	40·00	41·27	50·05	55·70	61 80
21	43·60	43·45	41·62	50·55	56·40	59·32
22	42·00	44·92	41·87	49·15	57·45	5!·35
23	40·45	44·30	43·65	·47·50	56·70	58·50
24	40·00	43·80	45·65	48·85	56·80	60·30
25	37·80	42·37	46·40	50·00	58·80	62·00
26	36·75	41·95	46·15	49·40	60·25	61·10
27	39·60	42·82	46·75	50·05	59·40	61·20
28	38·05	40·90	45·40	49·52	59·20	62·25
29	38·42		45·30	48·65	58·90	62·10
30	37·95		45·20	50·00	56·95	62·20
31	39·80		43·85		55·50	

in London, on an average of 10 Years: 1797—1806.

	Sev. Mo. July	Eight Mo. Aug.	Nin. Mo. Sept.	Ten. Mo. Oct.	Elev. Mo. Nov.	Twel.Mo. Dec.
1	61·90	67·40	61·05	54·10	48·85	42·30
2	63·40	65·80	61·25	55·90	48·80	41·35
3	63·45	66·50	62·35	56·40	45·75	40·30
4	64·55	65·30	62·35	54·90	44·65	38·85
5	63·00	64·83	61·85	53·65	43·35	42·00
6	62·15	65·00	61·95	54·30	44·90	42·80
7	62·55	64·55	61·35	53·95	44·55	39·95
8	64·15	65·05	60·10	54·70	46·30	40·30
9	64·35	65·60	60·00	54·00	45·15	39·95
10	63·45	65·67	59·60	52·20	44·80	39·35
11	63·55	65·60	59·65	52·45	46·10	39·35
12	63·05	66·35	58·25	50·55	44·60	38·85
13	62·35	64·10	58·90	50·20	44·60	38·60
14	62·60	64·30	60·35	51·05	45·85	38·40
15	61·90	65·00	61·30	51·55	45·70	39·20
16	62·85	64·35	60·62	50·75	44·80	40·40
17	63·70	65·95	61·20	50·00	43·05	39·95
18	63·05	65·45	61·10	52·05	42·40	39·15
19	63·80	65·00	60·70	51·10	41·95	39·35
20	63·90	63·80	60·20	50·00	41·40	40·65
21	62·75	62·55	58·20	50·95	41·55	40·75
22	62·85	62·35	56·80	49·90	41·50	41·35
23	64·45	62·95	57·25	48·45	41·50	40·25
24	63·60	63·05	55·60	48·80	40·90	39·00
25	62·70	62·05	54·75	48·60	41·90	38·70
26	64·25	61·00	55·55	50·30	42·95	39·15
27	64·05	62·10	57·15	48·45	41·90	38·45
28	63·50	64·25	54·75	47·50	42·35	35·50
29	63·95	65·10	56·85	47·10	42·00	38·90
30	65·35	64·15	55·15	48·50	42·00	40·20
31	66·05	62·65		49·15		37·75

MEAN TEMPERATURE of every day in the Year in

	First Mo. Jan.	Sec. Mo. Feb.	ThirdMo. Mar.	Four.Mo. April	Fifth Mo. May	Sixth Mo. June
1	36·00	39·75	42·50	44·00	54·50	58·10
2	33·95	40·30	41·85	43·25	52·85	59·00
3	33·50	39·80	40·70	42·35	54·05	57·25
4	34·65	38·50	42·00	42.80	53·00	58·90
5	35·65	40·35	39·00	44·55	53·05	56·00
6	36·05	42·25	40·45	46·80	54·05	55·60
7	34·70	39·35	40·20	46·25	54·10	57·45
8	34·90	39·25	40·80	44·65	55·25	59·15
9	35·95	39·10	40·15	46·30	54·00	57·15
10	36·60	40·85	38·10	47·90	54·25	57·60
11	36·40	42·35	40·50	45·80	56·00	58·80
12	34·90	41·00	40·25	47·40	55·55	58·55
13	34·30	39·55	40·45	47·25	56·65	59·80
14	33·10	38·95	39·05	47·60	55·90	59·00
15	31·10	40·15	38·65	46·80	56·50	57·50
16	33·20	40·25	39·30	46·15	57·15	57·75
17	32·05	37·65	39·95	44·20	58·75	58·00
18	32·05	37·60	40·75	44·30	55·65	59·00
19	32·35	38·70	41·45	46·55	54·65	59·00
20	30·75	38·35	44·35	48·15	55 15	59·15
21	31·10	39·95	45·25	47·00	54·25	59·65
22	30·70	40·30	43·70	48·20	52·30	58·35
23	32·70	38·85	42·65	49·05	54·45	58·75
24	33·20	40·65	40·90	49·10	56·05	58·85
25	32·40	39·95	39·30	49·15	58·25	61·10
26	35·65	39·50	41·75	49·30	58·45	59·20
27	36·05	39·95	43·90	50·35	57·60	59·95
28	35·30	39·97	46·00	50·90	58·60	59·45
29	34·85		44·95	49·40	57·85	61·30
30	35·75		44·15	51·15	58·50	60·60
31	38·90		44·60		60·45	

the Country, on an average of 10 Years : 1807—1816.

	Sev. Mo. July	Eight. Mo Aug.	Nin. Mo. Sept.	Ten. Mo. Oct.	Elev. Mo. Nov.	Twel.Mo. Dec.
1	60·25	62·15	59·05	51·60	47·15	39·90
2	60·85	64·10	59·55	51·60	45·95	39·00
3	57·15	62·00	59·80	53·60	43·05	40·95
4	56·80	62·15	57.50	54·95	41·90	40·95
5	59·65	62·10	58·85	56·60	41·30	39·40
6	61 35	61·75	56·40	54·80	41·90	39·40
7	61·45	61·85	55·55	53·60	41 30	37·70
8	62·00	60·90	55·65	52·90	42·25	36·15
9	63·40	59·30	57·10	51·25	44·30	35·75
10	62·25	61·70	56·55	51·90	42·65	36·45
11	64·60	61·10	57·15	51·20	42·70	37·05
12	65·00	62·35	54·60	49·65	43·10	39·25
13	64·75	61·45	54·90	51·05	41·10	38·55
14	64·00	62·25	56·05	50·65	40·65	38·00
15	63·30	62·25	57·35	49·90	41·50	38·15
16	61·90	60·95	57·45	50·15	41·20	36·95
17	62·65	61·10	56·70	51·20	41·00	39·05
18	62·00	62·90	56·85	50·60	39·25	39·55
19	63·95	60·45	55·05	51 10	38·55	37·45
20	62·60	60·05	56·70	51·55	40·85	35·70
21	61·00	60·75	57·85	50·30	39·00	33·60
22	62·00	62·65	58·60	50·05	37·80	35·40
23	64·05	62·90	54·80	47·55	38·55	37·20
24	63·80	60·55	55·20	48·15	38·70	36·75
25	65·70	61·90	53·80	47·15	40·65	36·45
26	63·10	60·35	55·60	46·20	40·10	37·65
27	62·95	60·60	53·85	46·55	38·10	35·05
28	63·20	58·85	52·00	45·10	36·95	36·70
29	63·65	57·80	53·70	44·95	37·80	37·80
30	61·80	61·75	53·20	45·85	38·90	37·20
31	61 15	60·80		46·10		37·25

MEAN TEMPERATURE of every day in the Year for

	First Mo. Jan.	Sec. Mo. Feb.	ThirdMo. Mar.	Four.Mo. April	Fifth Mo. May	Sixth Mo. June
1	36·57	39·70	42·27	44·17	52·75	57·05
2	35·92	40·37	42·80	44·37	52·55	57·85
3	35·60	40·32	42·10	43·87	52·67	58·15
4	36·42	38·34	42·10	44·82	52·77	59·22
5	37·47	39·20	39·69	45·67	53·22	57·72
6	37·12	39·47	40·22	46·84	54·57	57·45
7	35·85	37·37	39·54	47·10	54·70	58·70
8	36·05	37·27	40·05	46·72	55·07	59·52
9	36·12	39·05	39·65	47·17	54·20	59·15
10	36·07	39·92	38·90	48·32	53·87	59·37
11	35·62	40·00	40·60	46·60	54·74	58·75
12	34·45	38·37	40·72	46·76	54·22	58·40
13	35·27	38·10	40·47	46·57	54·12	59·75
14	36·20	37·42	40·90	47·44	53·47	59·67
15	35·05	39·22	40·81	48·16	54·35	58·55
16	35·65	38·90	40·51	47·95	55·30	59·12
17	35·52	37·82	41·27	47·00	56·65	59·55
18	36·12	37·92	41·75	47·22	55·32	60·17
19	37·02	38·37	41·25	48·52	55·07	59·77
20	36·92	39·17	42·81	49·10	55·42	60·47
21	37·35	41·70	43·44	48·77	55·32	59·49
22	36·35	42·61	42·79	48·67	54·87	58·85
23	36·57	41 57	43·15	48·27	55·57	58·62
24	36·60	42.22	43·27	48·97	56·42	59·57
25	35·10	41·16	42·85	49·57	58·52	61·55
26	36·20	40·72	43·95	49·35	59·35	60·15
27	37·82	41·39	45·32	50·20	58·50	60·57
28	36·67	40·44	45·70	50·21	58·90	60·85
29	36·64		45·12	49·02	58·37	61·70
30	36·85		44·67	50·57	57·72	61·40
31	39·35		44·22		57·97	

London and its environs, on an average of 20 Years : [1797—1816.

	Sev. Mo. July	Eigh. Mo. Aug.	Nin. Mo Sept.	Ten. Mo. Oct.	Elev. Mo. Nov.	Twel.Mo. Dec.
1	61·07	64·77	60·05	52·85	48·00	41·10
2	62·12	64·95	60·40	53·75	47·37	40·17
3	60·30	64·25	61·07	55·00	44·40	40·62
4	60·67	63·72	59·92	54·92	43·27	39·90
5	61·32	63·47	60·35	55·12	42·32	40·70
6	61·75	63·37	59·17	54·55	43·40	41·10
7	62·00	63·20	58·45	53·77	42·92	38·82
8	63·07	62·97	57·87	53·80	44·27	38·22
9	63·87	62·45	58·55	52·62	44·72	37·85
10	62·85	63·69	58·07	52·05	43·72	37·90
11	64·07	63·35	58·40	51·82	44·40	38·20
12	64·02	64·35	56·42	50·10	43·85	39·05
13	63·55	62·77	56·90	50·62	42·85	38·57
14	63·30	63·27	58·20	50·85	43·25	38·20
15	62·60	63·62	59·32	50·72	43·60	38·67
16	62·37	62·65	59·04	50·45	43·00	38·67
17	63·17	63·52	58·95	50·60	42·02	39·50
18	62·52	64·17	58·97	51·32	40·82	39·35
19	63·87	62·72	57·87	51·10	40·25	38·40
20	63·25	61·92	58·45	50·77	41·12	38·17
21	61·87	61·65	58·02	50·62	40·27	37·17
22	62·42	62·50	57·70	49·97	39·65	38.37
23	64·25	62·92	56·02	48·00	40·02	38·72
24	63·70	61·80	55·40	48·47	39·80	37·87
25	64·20	61·97	54·27	47·87	41·27	37·57
26	63·67	60·67	55·57	48·25	41 52	38·40
27	63·50	61·35	55 50	47·50	40·00	36·75
28	63·35	61 55	53·37	46·30	39·65	36·10
29	63·80	61·45	55·27	46·02	39·90	38·35
30	63·57	62·95	54 17	47·17	40·45	38·70
31	63·60	61·72		47·62		37·50

*Examples of opposite variations of the Mean Tempe-
rature in different periods.*

Before I dismiss the subject of Temperature, it will
be proper to exhibit specimens of the tendency of the
average diurnal Temperature to vary, at different periods, in opposite directions.

The upper curve in fig. 7 (the full line) is the diurnal
mean carried through the *First month,* on the average
from 1797 to 1806 : the corresponding Temperature,
on the average from 1807 to 1816, is represented by
the lower curve. The diurnal mean on each of the
above averages for the *Seventh month,* is shown by
the two curves in fig. 8. See the same in the Tables,
p. 138 — 141. From the contrast in direction which
prevails through the greatest part of these two figures,
it appears, that the Mean Temperature is subject to
these peculiar variations, both in the hottest and coldest months of the year. I shall revert hereafter to the
figures, and treat of the probable cause of the opposition. In the mean time the *dotted curves* may serve
to explain a discrepancy, betwixt my own diurnal mean
observations and those of the Royal Society, which,
not being obvious in the Tables of the latter, was not
detected till comparison had been fully made of the
respective results.

In the Register of the Royal Society, the *minimum*
by Six's Thermometer is that of the nocturnal depression *following* the maximum of the day indicated : in
my own it is that of the depression preceding it. In
averages relating to the month, or any longer period,
this difference induces no error in the comparison :
but when, as in the present curves, the mean of each
day is to be exhibited, there results a discrepancy, of
which it was not proper to leave the reader ignorant.
In the dotted curves, therefore, is shown the mean
Temperature by the Register of the Royal Society, as
it would have appeared, had it been calculated for
each day, according to my own method. The change
of the one curve for the other, it will be perceived, in
no way affects the contrast I have been insisting on.

OF THE PRESSURE.

NEXT to the Temperature, the variable *Pressure* of the Atmosphere, as indicated by the *Barometer*, claims the attention of those who are accustomed to the use of philosophical instruments. Indeed, the elegance of its construction, the facility of observing its changes, perhaps also something mysterious and imperfectly understood in its indications, have made this instrument but too successful a rival to the Thermometer : and we are probably deprived by this preference on the part of observers, of many useful results, which the latter, skilfully used might, in different situations, have furnished to science. The Barometer, when we contemplate it as a counterpoise to the weight of the atmosphere, is certainly a curious instrument : its movements, unlike those of the Thermometer, which relate only to surrounding space, bring us intelligence from the very surface of the aerial ocean, many miles above our heads. Here, probably, exist elevations and depressions of prodigious extent ; and as the representative in miniature of those tides in a sea without shores, its variations deserve, in point of theory, greater attention than has been hitherto bestowed on them : for the Barometer has been more observed than studied, and our knowledge of the principles on which its changes proceed is as yet little better than empirical. Hence frequent disappointment to those who trust to it

T

as a *weather-glass* — unless indeed it be assisted by attention to natural prognostics, and to the humbler, but not less certain indications of the Thermometer and the Vane. But our present business is, not so much with the theory of its movements, as with their *General results*; in which we shall find matter sufficiently interesting, and allied to the previous facts of this inquiry

Mean of the Barometer for London.

The mean height of the Barometer as deduced from 124 Lunar periods in this work, beginning Dec. 10, 1806, and ending Dec. 11, 1816, including a space of *ten years*, is .... 29.823 in.

The mean height for the ten years by the Calendar, from 1807 to 1816 inclusive, as deduced from the yearly results in the Philosophical Transactions, is 29.849

and for the ten years preceding 29.882

Average at London on 20 years ending with 1816 .... 29.8655

The Barometer employed at Somerset-house is uniformly stated, during this period, as situated 81 feet above the level of low water spring tides in the Thames. I am not prepared to state with equal precision the different heights (for the most part inferior to this) at which the observations contained in my Tables were made: nor is it of importance, as I do not propose in this instance to incorporate the results of that register with my own, for the purpose of drawing more extensive general inferences. The mean of 29.823 inches is therefore to be considered as the standard, to which I refer my own observations, for the ten years which are now to be more particularly examined.

Yearly Range and Extremes of the Barometer for Ten Years.

The General Table C exhibits the greatest and least elevations of the Barometer in each month, for the ten years from 1807 to 1816, together with the attendant winds. To the maximum heights of each year I have annexed the mark *, and to the minimum † The reader will perceive that the whole of the yearly *maxima* stand connected with *Northerly* winds, and the whole of the yearly *minima* with *Southerly*. Indeed, this rule holds generally throughout the Table, as to the monthly extremes also; and I need scarcely refer in this place to the fact, so long known and proved, that Northerly winds raise the Barometer, while Southerly ones depress it.

Of the yearly *maxima*, the greater number occur within the first three months of the year, and the rest about the end of it. The yearly *minima*, with a single exception, fall within the last three months. Thus there are six months, of Spring and Summer, in which, with a single exception in ten years, the Barometer visits neither extreme of its yearly variation: while the higher annual extreme is chiefly the product of Winter, and the lower one of Autumn.

The following Table drawn from the results of Table C will serve for more easy reference.

Year.	Mean of 12 greatest elevations.	Mean of 12 greatest depressions.	Medium of elevations & depress.	Highest observation in the year	Lowest observation in the year	Range for the year.
1807	30·310	29·167	29·738	30·60	28·68	1·92 in.
1808	30·338	29·263	29 800	30·71	28·72	1·99 in.
1809	30·295	29·088	29·691	30·49	28·25	2·24 in.
1810	30·323	29·327	29·825	30 51	28·50	2·01 in.
1811	30·302	29·195	29·748	30·61	28 65	1·96 in.
1812	30·266	29·279	29·772	30·51	28·53	1·98 in.
1813	30·314	29·214	29 764	30 50	28·64	1·86 in.
1814	30·266	29·190	29 728	30 42	28 22	2·20 in.
1815	30·309	29·136	29·722	30 58	28 85	1·73 in.
1816	30·327	29·023	29 675	30 62	28·53	2 09 in.
Averages	30·305	29·188	29·746	30·555	28·557	1·998 in.

The average of the third column, or the *medium* between the average elevations and depressions, is near eight hundredths of an inch below the *mean height for the climate* (or 29.823 in.), the reason of which is, that the depressions occupy a smaller space of time than the elevations; in consequence, a less proportion of them comes into an average founded on daily results.

The *average annual range* for 10 years is very nearly 2 inches; the range varies in different years about ½ an inch.

The *greatest elevation* in 10 years appears to have been 30.71 inches. This took place on the 24th of Second month, Feb. 1808: it was introduced by NE breezes, with hoar frosts at night, and a temperature of 39° in the middle of the day But this is not quite the higher extreme of the climate: for on the 7th of the same month in 1798, the Barometer rose to 30.89 inches: the elevation being in like manner introduced by a gentle NE wind, with hoar frosts at night, and a temperature of 39° in the middle of the day on which it took place. I observed on this occasion, that the air at Plaistow was filled with a dense mist : but in 1808 I believe it was clear at the time. The coincidence of circumstances in some other points is remarkable.

The *greatest depression* in 10 years occurred in 1814, on the 29th of the First month, Jan. when the Barometer descended to 28.22 in. It is very nearly equalled by a former one, on the 17th of Twelfth month, Dec. 1809, which was 28.25 in. Both were introduced by strong Southerly winds. Having been on each of these occasions at home, and attentive to the phenomena, I must refer the reader to the accounts of them, in the Notes and Results under Table 39, and the Results under Table 90, in my first volume. The depression of 1814, it will be observed, took place at the first remission of the severe cold of that season, by which the Thames was frozen over.

Neither extreme for the year is ever produced very suddenly. In 1798, the Barometer took eight days, to rise from 29.15 to 30.89: in 1814, five days, to fall from 29.88 to 28.22. The great depression of 1809 was in progress for several weeks, before it arrived at the crisis : but of this I shall have occasion to treat hereafter. There is also, as in the case of Temperature, a consistency between the annual extremes : in those years in which the Barometer falls very low, it does not rise so high as in others, and *vice versa:* the same sort of gradation from year to year which appears in the Temperature, is also occasionally found in these results.

Monthly Range and Extremes of the Barometer on an average of Ten years.

This part of the subject presents gradations almost as regular and striking as those of the Monthly Temperature : to exhibit which it will be necessary to have recourse to curves, and to a second statement in figures, drawn from the General Table C. The present Table consists entirely of results found on taking the columns vertically by the month; as the other did of those found on taking the lines horizontally for the year.

Month.	Average of Maxima.	Average of Minima.	Difference or mean Range.	Greatest elevation in 10 years.	Greatest depression in 10 years.	Difference or full Range.
1. Jan.	30·400	28·971	1·429	30·60	28·22	2·38 in.
2. Feb.	30·419	29·069	1·350	30·71	28·70	2·01 in.
3. Mar.	30·405	29·106	1·299	30·61	28·81	1·80 in.
4. Apr.	30·233	29·154	1·079	30·36	28·74	1·62 in.
5. May	30·251	29·337	0·914	30·42	28·90	1·52 in.
6. June	30·283	29·452	0·831	30·40	29·15	1·25 in.
7. July	30·182	29·491	0·691	30·39	29·40	0·99 in.
8. Aug.	30·193	29·434	0·759	30·26	29·24	1·02 in.
9. Sept.	30·232	29·334	0·898	30·40	28·86	1 54 in.
10. Oct.	30·212	29 056	1·156	30·35	28·53	1·82 in.
11. Nov.	30·357	28·899	1·458	30·62	28·50	2·12 in.
12. Dec.	30·407	28·957	1 450	30·62	28·25	2·37 in.

The upper curve in Fig. 9 shews the manner of variation throughout the year of the average higher extreme, or mean of ten maxima; the lower curve that of the average lower extreme, or mean of ten minima. I have added a medium curve between the two; together with a set of dotted perpendiculars, which express the *mean range* for each month.

The greatest *elevations* of the Barometer, it appears, take place in the winter months; and during seven months of the year, from the Fourth to the Tenth inclusive, they fall off, to the amount, as the fourth column shews, of a quarter of an inch on the whole.

Fig. 9.

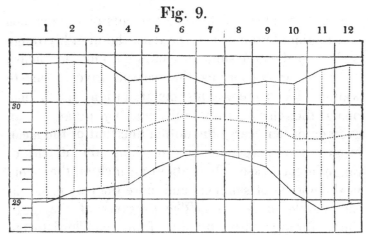

The *depressions* too, are most considerable in the winter half year, being at their full extent in the Eleventh month : from whence they decrease to the Seventh and then increase again ; the total difference exceeding half an inch : the progress of the series through the six months of summer forms a regular curve, ascending and descending ; the remainder of it an ascending line.

Fig. 10.

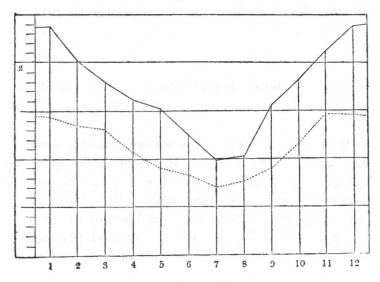

In consequence of these movements of the maxima and minima in opposite directions, the monthly range is shortened as the summer comes on, and lengthened again in proportion as that season gives place to winter Fig. 10 represents this gradation ; the full line being the curve of the full monthly range on an average of ten years, the dotted line that of the mean range ; the perpendiculars marking the extent of range in each case. It is about an inch on the whole in the middle of summer, and more than $2\frac{1}{3}$ inches in the middle of winter. The progress of the mean range through eight months, from the Third to the Eleventh, forms again a regular curve, in descent and ascent ; the remainder a descending line.

From the greater tendency of the *depressions* to go off in the summer, the medium curve has its higher points in that season, notwithstanding the lower level of the great *elevations*. In this respect, as will be shewn hereafter, the curve of the *medium* agrees with that of the true *mean*, deduced from all the observations ; and each of them proves that on the whole, *the weight of the atmosphere is greater in summer than in winter.*

Remarks on the Scale of the Barometer, and on the use of this instrument as a Weather-glass.

It was not without reason, we see, that the scale of the Barometer for London was fixed in the space of 3 inches, from 28 in. at the lower, to 31 in. at the higher extreme. The same scale will probably serve for every place, situate near the level of the sea, in these islands. But for inland elevated places, the scale should be brought down half an inch, or in some cases an inch, making its lower extreme $27\frac{1}{2}$ or 27 in., while the higher part might be shortened in proportion. For it cannot be doubted that in some of these, the quicksilver would occasional y fall below the bottom of the scale of a well adjusted Barometer; and a false adjustment, by means of the screw at bottom, or purposely effected in the construction, is probably the reason why this defect is not more often perceived.

The terms *Fair, Changeable, Rain,* &c. at present commonly attached to certain points of the scale, are likewise misplaced, as far as regards London and other places near the sea-level. The true *medium* of elevations and depressions for these, appears to be very near to 29·75 in. This part of the scale therefore, and not 29·50, as at present, should be marked *Changeable.* Half an inch above it, or 30 25, may be designated *Fair*; and the same distance below, or 29·25, *Stormy.* These limits might then be of some real use to the inexperienced observer; especially if his judgment were assisted by some such directions as the following.

Rain is most plentiful and Thunder most frequent, while the quicksilver fluctuates *about the changeable point,* or between that and the stormy one; for the mean or average of these short movements being generally lower in winter, and higher in summer, an allowance of one or two tenths may be made either way, according to the season.

In proportion as the quicksilver advances from the changeable towards the *Fair* point, the probability of fair weather increases : at or beyond that point, it is extreme : and it decreases as the quicksilver recedes again.

In proportion as the quicksilver falls from the changeable towards the *Stormy* point, the probability of a storm of wind increases: at or below that point, it is extreme: and the rising again of the quicksilver is not to be regarded as indicative of more settled weather, until it has again passed the changeable point.

Sudden considerable changes, in either direction, are commonly followed by fair or foul weather equally transient: while a steady rise from day to day, of a tenth or so in the twenty-four hours, or a prolonged fall in the same proportion, (either of them passing the changeable point,) may generally be trusted as prognostics of continued rain or fair weather.

It is obvious, that in situations exceeding 200 feet in elevation above the sea level, the changeable point may very well remain at 29·50; and that in those exceeding 500, it should be lowered at least to 29·25, the others in each case being placed at proportionate distances. Indeed, to render the Barometer as useful as it is capable of being, in different respects, every separate instrument should be adjusted by the maker to some good common standard; and the *mean point* fixed with an attention to the known elevation of the country, similar to that which is exercised with regard to the latitude in constructing a dial.

So much as to this instrument when employed, by itself, as a Weather-glass; but it will always be found most useful in this respect, when observed in conjunction with the wind and temperature, and with the natural prognostics daily presented to our notice on the face of the sky.

OF THE WINDS.

THE direction in which the wind at any time passes over us is far from constituting the whole of what we would wish to know on this subject. The length and breadth of the stream, its mean depth and velocity; the part of it in which we are at any time situate; the place where it took its origin, and that in which it wheels about to assume a new direction, or having spent its force, becomes stagnant : all these are objects of reasonable curiosity, and which might perhaps be ascertained by distant and well-concerted observations. At present, we are able to infer only now and then a consequence, from the comparison of results found at home with those deducible from other registers, or from the reports obtained from the coasts by mercantile men; who are sometimes deeply interested in the cessation or continuance of particular winds.

The *Yearly* and *Monthly* results of the observations contained in the first volume of this work, will constitute the matter of the present section. In digesting these, I have assumed *five* classes for the winds. The observations would have furnished *nine*, but with limits less entitled to confidence than those which we obtain by embracing a greater number of points, and thus giving a chance of mutual compensation to some inaccuracies, inseparable from the smaller divisions.

The 1st Class extends from *North* to East, not including the latter point; that is, it consists of my N and NE observations : and so of the rest.

The 2d Class extends from *East* to South, not including the latter

The 3d Class from *South* to West, the latter not
included.

The 4th Class from *West* to North, (not included,)
completing the compass.

The 5th Class comprehends the *variable* observations.

Yearly proportions of the several Winds in Ten years.

The following Table contains a statement of these
on the plan which I have described, the few days
wanting in my Tables being supplied, for the purpose
of calculation, from the Register of the Royal Society.

Year	N—E	E—S	S—W	W—N	Var.
1807	69	34	113	114	35
1808	82	38	108	103	35
1809	68	50	123	91	33
1810	81	72	78	83	51
1811	58	59	119	93	36
1812	82	66	93	91	34
1813	76	53	92	124	20
1814	96	65	91	96	17
1815	68	36	121	107	33
1816	64	66	106	102	30
Averages	74,4	53,9	104,4	100,4	32,4

This Table shews that, with some variation in different years, there obtains a proportion between the
different classes of winds in our climate, which may
be thus stated.

1. A wind from the North, or between that and
East, prevails on an average 74 out of 365 days; the
greatest amount of its number being 96, the least 58
days.

2. A wind from East to South, 54 days, varying in different years from 72 to 34 days.

3. A wind from South to West, 104 days, varying from 123 to 78 days.

4. A wind from West to North, 100 days, varying from 124 to 83 days.

5. Variable winds obtain about 33 days, or the remainder of the year, their number being from 51 to 17 days,

The last mentioned division, from the arbitrary manner of noting, is probably the least exact in its limits : there being undoubtedly many days on which the observation might have been carried to one of the four classes, as *prevalent*; and others, on which the term *variable* might have been applied, in preference to the denomination set down. Yet amidst this uncertainty, it is worthy of remark that in *seven* out of the ten years its proportion varies only from 30 to 36, which would induce the conclusion that, were the observations uniformly attended to in this respect, the days on which the wind changes with some force *to an opposite point in the course of the day,* would be found between those limits.

If we now make of the whole two great divisions, towards East and West, allotting the variable to each in due proportion, we shall have

$$\begin{array}{ll}
\text{Easterly winds} & . \ . \ 140 \\
\text{Westerly} \quad . \ . & . \ 225 \\
\hline
& 365
\end{array}$$

If towards North and South, then,

$$\begin{array}{ll}
\text{Northerly winds} & . \ . \ 192 \\
\text{Southerly} \ . \ . \ . \ . \ . & 173 \\
\hline
& 365
\end{array}$$

Thus a Westerly direction is found to preponderate by about a *third* over the Easterly; and a Northerly direction by about a *ninth* over the Southerly, in the winds of these ten years.

I suppose that a careful revision of the observations, with the aid which might be got from other registers, would introduce some corrections, but probably not any alterations of moment, into these averages. A different series of years in the same district of the island, or the same series in a different district, might also give some variation in the results. The reader is therefore to be on his guard against applying them *generally*, at least for the present. I have no comparative results to introduce on this occasion.

Monthly proportions of the different Winds for Ten years.

The following Table exhibits these in days and decimal parts, the classes being as before, and the term from 1807 to 1816.

Month	N—E	E—S	S—W	W—N	Var.
1. Jan.	6,8	5,3	7,0	9,1	2,8
2. Feb.	3,2	4,0	11,7	7,4	1,7
3. Mar.	9,8	5,4	6,6	6,5	2,7
4. Apr.	8,3	5,6	6,0	6,4	3,7
5. May	5,9	6,5	9,0	5,6	4,0
6. June	7,1	3,0	7,2	9,1	3,6
7. July	4,5	2,5	9,5	11,5	3,0
8. Aug.	3,5	2,9	10,2	12,9	1,5
9. Sept.	6,4	6,0	8,0	7,4	2,2
10. Oct.	5,2	5,0	10,5	7,4	2,9
11. Nov.	7,8	3,1	8,8	8,4	1,9
12. Dec.	5,0	4,6	9,9	9,7	1,8
Averages	6,00	4,50	8,70	8,45	2,65

In the *First Month,* which may be regarded as the middle of winter, we have little more than a mean proportion of N—E winds : yet the *Northerly,* taken together, preponderate by a *fourth* of their amount over the Southerly winds.

In the *Second Month,* the proportions of Northerly and Southerly are reversed, the latter exceeding the former by a *third :* and this principally through the falling off of the N—E to one half, and the increase of the S—W to their highest proportion for the year.

In the *Third Month,* the N—E are in greater proportion than in any other part of the year, exceeding their own average by more than a third.

In the *Fourth Month,* the N—E winds abate somewhat of their excess, continuing still in very high proportion. This and the preceding month exhibit about the same total preponderance of Northerly winds, as the First month : and in both, the E—S class being above its average, the general Easterly direction prevails over the Westerly.

In the *Fifth Month,* the Southerly winds resume the like superiority as in the Second. The E—S class is at its maximum. The N—E having decreased for two months, is now below its average : and the W—N which has decreased by an uninterrupted gradation from the First month, is at its *minimum* proportion : the *variable* winds are at their highest amount.

Sixth Month : a preponderance of *Northerly* winds by more than a *third ;* chiefly from the return of the W—N class.

Seventh Month. In this month, the class of W—N decidedly prevails over the rest : the S—W is also in high proportion ; the N—E very low, and the E—S at its *minimum,* having gone off for two months.

The *Eighth Month* exhibits the class N—E at its *minimum*, and that of E—S but little removed from it: while the W—N is at its *maximum*, having increased for three months, and the S—W in high proportion, having increased for two months. This month has the least proportion of variable winds.

Ninth Month. We have here almost a balance between the Northerly and Southerly winds. In other respects, the class E—S, (which we must remember comprehends the former *point* and excludes the latter), takes a little from the rest, and is but little short of its highest amount.

In the *Tenth Month*, the winds on the North and South sides of East are very nearly equal: but the S—W class predominates over the whole, and with the aid of the E—S, exceeds the Northerly winds by a fourth of the sum of the latter.

Eleventh Month. Northerly winds now predominate by a fourth of their amount; chiefly from the increase of the class N—E; and the proportion of *variable* is very small.

Twelfth Month. The classes in this month do not depart very far from their respective averages. We have again the Northerly and Southerly almost exactly balanced; while the Westerly are nearly double the sum of the Easterly.

The monthly proportions of the several classes *in each year*, will be found in the General Table D, over the Monthly amounts of rain. I shall have occasion to resume the subject more than once, in the course of this volume, with a view to the connexion of particular winds with the variations of the Barometer, or with dry and wet seasons; and their relation to the Lunar periods, the Solstices, and the Equinoxes; but it is proper first to present the reader with the imme-

diate Results of the Register through its several divisions.

The subject of the Winds is one of so great interest
to the community, that nothing but the apparent want
of system, in their variations in these latitudes, can
have prevented men of science from studying them
with greater attention, and bringing out some useful
results. I believe the experience of our navigators,
in this as in some other respects, outruns science, and
furnishes already some general axioms, respecting the
Winds commonly met with at particular seasons in
our climate. It would be rendering no small service
to those who have frequent occasion to quit our coast,
or to enter our harbours from the seas, could the whole
of the information already within our reach on this
subject be digested in a systematic form for their use :
more especially, as it might enable them to anticipate
with greater certainty the recurrence of those long
periods of NE and SW winds, not improperly termed
the Monsoons of our climate ; by which our communication with the Atlantic is at times impeded, at others
facilitated, for whole months together.

x

OF THE EVAPORATION.

Experiments made with a view to ascertain the natural Evaporation differ in their results according to the manner in which the water is exposed. If it be fully acted upon by the sun and wind, in a vessel of small capacity, the quantity evaporated appears in excess : if greatly sheltered from both, the contrary. In the whole of the experiments detailed in the First volume, the water was placed under cover; just sufficiently sheltered to prevent the entrance of driving rains, and consequently the direct impulse of the sun's rays when much elevated above the horizon. During the first three years, the results were entered almost daily in the Tables; afterwards, at intervals varying from two to ten days, or weekly. But in the year 1815, having substituted for these results the daily indications of the Hygrometer, I ceased to attend so constantly to the Evaporation.

Mean Evaporation in the Year.

In the years 1807, 1808, and 1809, the guage being elevated about 43 feet from the ground, exposed to the SE, and subject to the free action of the wind in most directions, the annual average result was 37.85 inches.

From 1810 to 1812 inclusive, the instrument being in various situations, for the most part lower and less exposed, the annual average was 33.37 inches.

Lastly, in the space from 1813 to 1815 inclusive, the guage being upon or near the ground, the annual results averaged only 20.28 inches.

Having resumed in 1818 the observations conducted on the ground, I obtained in eleven months a total of about 23 inches : to which if we add a mean result for the month omitted we shall have for this year (so remarkable for its hot and dry summer) an evaporation of nearly *twenty-five* inches : a quantity not disproportionate to the average of the three years last recited, considering that the calmness of the air during this summer was as remarkable as its temperature.

The Evaporation obtained in 1818 very nearly equals the annual average depth of rain about London. In years with a cool or wet summer it falls below this standard : but on account of the acceleration to which this process is liable by the effect of strong winds, it is difficult to make an accurate comparison in two seasons of unequal temperature. Nor is it likely that with a little water, exposed in a vessel of a few inches diameter, we should obtain a complete solution of this problem, as it is set before us on the great scale of nature. The first or second of the three averages above stated may perhaps approximate to the Evaporation from our rivers, the surface of which is always in motion by the winds and currents : the third may be considered as representing that of small canals, ponds, and reservoirs.

Evaporation in the different Seasons of the Year.

The monthly results which form the basis of this section are digested in the general Table E, at the end of the volume. The series (of eight years and a half) is an interrupted one, but it presents at least

x 2

seven results for constructing each of the monthly averages at the foot of the Table. These averages run, as might have been expected, with a general, but not uniform, relation to the Monthly Mean Temperature. The cause of Evaporation is the heat contained in the fluid, and it has been long since shewn that, other things being equal, the effect is in relative proportion to the temperature. But in nature it is always modified by the quantity of vapour already subsisting in the atmosphere, considered relatively to the temperature of the latter. For instance, in the Third month 1807, the mean temperature of which was 42°, the Evaporation amounted to 2·66 in. : but in the Tenth month of the same year, with a temperature somewhat exceeding 42°, and more wind, it was only 1·86 in. : the difference being plainly caused by the season.

If we take the Twelfth, First, and Second months as the Winter, and the remaining months in similar classes of three, for the other seasons, and divide the average Evaporation among the four classes, it will stand thus :

	Evaporation.	Mean Temp.
Winter . .	3·587	37·20
Spring .	8·856	48·06
Summer	11·580	60·80
Autumn . . .	6·444	49·13

I have added the mean Temperature for each season as thus divided, the total of degrees of mean Temperature being 195·19, and that of mean Evaporation 30·467 for the year.

Then 195·19 : 30·467 : : 37·20 : 5·806
195·19 : 30·467 : : 48·06 : 7·501
195·19 : 30·467 : : 60·80 : 9·490
195·19 : 30·467 : : 49·13 : 7·668

The four results thus brought out being the quantities which ought to have been raised in vapour in each season, had the effect been in strict proportion to the Temperature, it follows, that in the three months here taken as *Spring* the Evaporation is *augmented* by about a sixth part, and in those taken as *Summer* above one fourth part, in consequence of the *dryness* of the air in these seasons : while in the three months taken as *Autumn* it is *lessened* by more than a sixth, and in those taken as *Winter* by considerably more than a third, in consequence of the *dampness* of the air.

To examine more particularly the monthly results—we see that, as the Temperature advances in the fore part of the year, the Evaporation on the whole increases steadily ; but in particular years it receives a check in some part of the spring, which is afterwards made up by a sudden increase. The reason of this is sometimes obvious in the variations of Temperature ; as in the year 1809, where I have annexed the mean Temperatures to the results. The rate is likewise occasionally kept down in this part of the year (as in the latter months,) by frosty weather. The very great increase in a fine spring may possibly be due, in part, to the electric state of the air in such seasons. For although electricity, in the low degree in which it is applied by nature at the Earth's surface, may not sensibly promote the actual emission of vapour from water, it may tend greatly to increase the retentive power of the air, by rendering the particles of the mixture of gases and water in a higher degree mutually repulsive, or in other words, by keeping up the *elasticity* of the atmosphere.

I have attributed an occasional low rate of Evaporation in Spring to the state of the Temperature.

Without destroying this position, we may however invert the term, and say that it is then even colder *because of the evaporation.* It cannot be doubted that the *sharpness* of our NE breezes in Spring is in great measure the result of their excessive dryness, relatively to the Temperature which prevails: in consequence of which they abstract the heat from the animal system by means of the moisture on the skin, which they convert into vapour with peculiar rapidity.

In the latter part of Spring the *guage* sometimes indicates an abundant supply of vapour, when in fact very little is poured into the atmosphere from the Earth; the surface, and even some considerable depth under it, being already dried by the sun and wind. It is then that we perceive the effects of the natural irrigation, carried on by means of the vapour diffused in the day time from canals, rivers, &c. and condensed by night in copious dews, which descend on the neighbouring herbage. Should the season afterwards prove showery, a great quantity of the first water that falls is vapourised by the heated Earth, with a rapidity of which, again, the *guage* gives no proper indication. This vapour may even continue to be thrown up, after the air has begun to approach towards saturation, and thus contribute to the formation of the next rain. And the water may be thus driven from the Earth to the clouds, and returned again in rain, until the surface, being cooled down, is prepared for desiccation under the solar rays by a drier current. The sudden change from a dry to an extremely humid state of the air, immediately after our Spring and Summer showers, is often sufficiently obvious to be detected by the most superficial observer: it is generally due to this sudden and copious production of vapour at the surface. The Spring and Summer are our most variable seasons in point of hygroscopic dryness.

In the Autumn, or rather at the approach of Winter, the rate of the production of vapour declines with great rapidity. The commencement of a saturated state of the air, while as yet precipitation has not generally commenced, gives to our fine autumnal weather a delicious *softness*, the reverse and the compensation of those keen blasts which so often attend the vernal season.

But this state does not continue long. On the approach of the first frost—indeed during a great part of our ordinary winters, the earth and waters retaining a temperature somewhat above that of the air, continue by the force of this inherent warmth to emit vapour. This is continually undergoing decomposition, and it fills the air with a *mist*, which, when by no means dense enough to constitute what we call *fog*, would yet appear to an observer stationed above its limits, as a white veil thrown over the whole face of the country : thicker indeed in the valleys and along the course of the rivers, but nowhere in our district surmounted by the land.

Evaporation under different circumstances of Wind, Temperature, &c.

There are few *days* in the whole year in which some vapour is not raised from the guage : but the process is apparently suspended while *dew* falls by night. A state of the air analogous to this appears to be the cause of its complete interruption by day : of which the reader will find some instances in the Tables, chiefly in the vicinity of the Winter solstice, and at the approach of frost.

It is not always suspended during *rain,* as I have ascertained by direct experiment. The rate is however usually much less on those days in which rain falls, and it is liable to a rapid increase immediately afterwards.

Sometimes, an excess in the rate is found to *precede* rain, whether from the agitation of the air, or the effects of electricity, or from both causes, I have not attempted to determine.

The calm which attends a change of wind sensibly lowers the rate: which also decreases, as might be expected, upon the going off of a wind which has blown steadily for some days.

A moist current of air flowing in upon us will sometimes check the Evaporation, although rain be not produced from it.

Examples of the gradual increase of the rate of Evaporation, in consequence of an elevation of the daily Temperature, and its decrease by the contrary, as likewise of the variations to which it is subject in windy weather, are so numerous in the Tables that it is needless to instance them.

Greatest and least Monthly and Daily Evaporation.

The greatest Evaporation in a month *by the higher guage,* was about 6 inches: in this case a number of favourable circumstances appeared to concur: a high Spring Temperature succeeding to protracted cold; dry winds and an abundant electricity.

The smallest monthly results are found at the approach and during the continuance of the great frost of 1813-14. And here we have a striking example of the retarding effect of a moist air on the process. In the last month of 1813, with great *fogs* prevailing, the mean Temperature being 38° 43, the Evaporation was 0·21 in. (in all probability the lowest amount in ten years,) but in the first three months of 1814, the frost having set in with rigour, and cleared the air, we have a gradation of increasing results thus, 0·25, 0·36, 0·83 in. with the mean Temperatures 26° 71, 33° 17, and 37° 82, all inferior to the former, and the first of them almost 12° below it: which difference in the effect is plainly due to the extreme dryness of the currents prevailing in the latter period.

Indeed, the most intense *cold* is insufficient of itself to put a stop to the formation of vapour. Ice evaporates freely during a clear frosty night; as I have repeatedly convinced myself by direct experiment: see Tables 28, 90, 95, of the first volume. In the former of these experiments, a circular area, five inches in diameter, lost 150 grains between sunset and sunrise. This is at the rate of more than 8000 Troy pounds of ice, or near 1000 gallons of water, from an acre of surface, in that time. The absorption of heat, necessary to the composition of so much vapour within a small space of the atmosphere, must be prodigious. In one instance of this kind, I found the depression of Temperature to exceed 10 degrees; the Thermometer on the snow being at 6° 5, while that at five feet elevation was at 17°. Some part of this cold however, might be ascribed to the radiation from the surface of the snow.

With these facts before us, we need not wonder to hear that a moderate fall of snow is sometimes entirely taken up again, during a succeeding Northerly gale, without the least sign of liquefaction on the surface. In deeper snows, the surface after a while becomes curiously grooved, scooped, and channelled, from the same cause: which effect is most conspicuous around

Y

the trunks of trees, and near the interstices of paling,
—in short wherever the stream of air acquires force
in a particular direction. A little observation will
satisfy any one that the snow is not removed, on these
occasions, merely by being driven before the wind.

Consistently with these facts, a sensible change in
the air to a dry state, after damp foggy weather in
winter, may be always safely placed among the indications of approaching frost.

To return to some considerations connected with
the higher extreme of Temperature—a rapid decrease
of the daily Evaporation in hot weather might furnish
a prognostic of approaching thunder and rain, were
it needful to add to those we possess already. The
greatest Evaporation in one day (a single instance excepted) which I have ever seen, occurred on the 17th
of the Fifth month, May, 1809. On that day the
amount was 0·39, on the following day 0·28, on the
next 0·14 inches : the corresponding mean Temperatures being 67°, 70° 5, and 64°, and consequently
furnishing, in respect of heat, no adequate cause for
the decrease. But in the evening of the 19th occurred
that tremendous storm of hail, rain, and thunder,
which I have particularly described under Table 32
of the first volume : and I cannot help supposing that,
on this occasion, the local influence of heat, aided by
an electric charge in the air, had suddenly raised, as
it were, a mound of vapour into those elevated regions, which it rarely visits in these latitudes, and
where it is subject, from the contiguity of an intensely
cold medium, to complete an extensive decomposition ; in which seems to lie the true cause of the prodigious developement of electricity manifested on
those occasions. In the same Table, the reader will
find a decrease of the daily Evaporation in this ratio,

0·33, 0·26, 0·19, followed by a tempest of wind, and a week's wet weather: but in this case the Barometer, Temperature, and Sky, furnished concurrent indications.

The going off of the excessive heat in the Seventh month 1808, of which I have already treated, p. 110, &c., was attended at Plaistow (although the reaction in the atmosphere took place about Gloucester) with the following rapid decrease in the daily rate of Evaporation. The hottest day being 0·35, the four following were 0·31, 0·27, 0·20, 0·16 inches: and this without more than a few drops of rain in our own immediate neighbourhood.

OF THE MOISTURE BY THE HYGROMETER.

While the Evaporation Guage indicates the rate of the production of vapour from surfaces capable of affording it to the air, the Hygrometer informs us of the state of the latter as approaching more or less towards a relative maximum of moisture, the existence of which, whether in the higher or the lower atmosphere, is commonly followed by rain.

At elevations of a few feet, the index of the hygrometer is usually found at sunrise on the moist side of the mean for the season. As the sun advances, in a fine morning, it recedes towards the dry side, sometimes with considerable rapidity, passing through twenty degrees of the scale by noon. In the evening, if not earlier, it returns again towards moisture. To obtain a true *mean* of variations so considerable, it would be needful to take a number of observations at equal intervals through the twenty-four hours, and average them. But I am not aware that any observer has yet gone so far as to obtain the extremes indicated in that period, in order to record a daily *medium*. In fact the present observations have the same disadvantage as would attend those on Temperature, were the Thermometer inspected but once in the day at a fixed hour. The time which mere convenience induced me to adopt is nine in the morning. About the Equinoxes, this hour is a medium between sunrise and noon, and consequently a very fit time to obtain a mean result. But in winter it approaches too near to sunrise, and in summer recedes too far from it. Imperfect as my results are, on this and other accounts, they are yet too valuable to be passed over, and I shall here give a summary of them with some remarks.

Monthly mean of De Luc's Hygrometer for four years, from daily observations made at 9 in the morning.

	1815	1816	1817	1818	1819	Averages
1. Jan.		81	80	78	81	80
2. Feb.		78	67	81	75	75
3. Mar.		67	64	65	71	67
4. April		59	52	61	69	60
5. May		†57	52	†57	64	57
6. June		†49	47	†47	66	52
7. July			47	45	63	52
8. Aug.	†50		52	†47	59	52
9. Sept.	†61	64	58	†66	71	64
10. Oct.	73	78	†59	73		71
11. Nov.	79	84	†76	81		80
12. Dec.	79	83	78	80		80

The results marked † are the mean of a deficient number of daily observations, varying from thirteen to twenty-five days in the month.

The *general mean* of these observations is 66 : and the state of moisture about the Equinoxes, it will be observed, approaches near to the mean, the average of the Third month being one degree moister, and that of the Ninth two degrees drier.

The extreme monthly averages are 80 for Winter, and 52 for Summer, which points are equidistant from the general mean. In the Spring and Autumn, the averages exhibit a gradation from each of these points to the other. It is obvious that had the observations been made always at noon, the medium and extremes would have been respectively, and perhaps proportionately, nearer to dryness : had they been made always at an hour equidistant between sunrise and noon, the two extremes would have approximated nearer to the medium. As it is, they mark strongly the character of the respective seasons of our climate in point of moisture or dryness; and those of our medical practitioners who at present attend to the Thermometer, as an assistant to their judgment in anticipating the prevailing diseases of the season, will perhaps be able, with the help of some such standard as this Table, to avail themselves of the Hygrometer also, for the like laudable purpose.

Connexion of the movements of the Hygrometer with Evaporation and Rain.

In general, a comparative degree of dryness by the Hygrometer is connected with Evaporation and fair weather; and of moisture, with precipitation and rain; regard being had, in both cases, to the *mean* of the season. During the Lunar period, commencing with the 8th of the Fourth month 1817 (Tab. 130), there fell only 0.28 in. of rain: the Hygrometer at nine a. m. was never beyond 63, and once at 34, the mean of the period 49. In the next Lunar period, there fell 3.18 inches: during which the Hygrometer was once at 80, and the mean of the period was 54.

But there are exceptions to be noticed here. In Summer, when precipitation is actually going on above, and thunder-clouds are already formed, the air below may continue, from the intense heat and the arid state of the soil, hygroscopically dry. An instance of this occurs presently after the above, in Tab. 132, Sixth month 24, Hygrometer at nine, 47, (the mean of the period,) " Morning cloudy, then fine: in the evening, heavy rain, with hail, thunder and lightning: Hygrometer before the storm, 36." Next morning at nine it was 61, the rain having afforded vapour: but dry indications again came on, followed by other thunder-storms, to the close of the period, which afforded 2.81 inches of rain.

In Winter, on the contrary, the air is sometimes very moist for a considerable time, without rain: chiefly during the prevalence of foggy days and frosty nights, with a high Barometer: for an instance of which see a space of more than two weeks, following the Winter Solstice, 1818, (Tab. 151). In clear sunny days of frost, however severe, it is otherwise: the Hygrometer indicated several times a dryness of 46 to 50 in the middle of the day during the intense cold in the Second month 1816. See Tab. 115, which also presents some examples of the moist extreme, followed by snow. In Tables 124, 125, 126, (in the Autumn and Winter of 1816) will be found a number of examples of hibernal moisture, some with and others without the accompaniment of rain. For an example of Summer moisture brought on by the fall of rain, it may suffice to point out Tab. 145, Seventh month 12; where the Hygrometer, having been for several mornings at 46—42, was brought at once by this cause to 70.

A rapid movement of the index towards dry in the morning seems to indicate a fair day, notwithstanding unfavourable signs in other respects. See Tab. 116, Third month 16: also Tab. 110, Eighth month 29, consulting the Notes. But extreme and unusual dryness should be suspected: see the same Table, under Seventh month 14, when a kind of *Harmattan* seems to have been blowing, and the index receded to 22, (the driest point at which I have seen it) yet rain followed in about 48 hours, though a very dry time before, and for some days after.

On the other hand, if the index, when found on

the moist side, in the morning after a fair day, continue stationary, or advance to a higher number, rain is to be expected; and this is perhaps among the most certain indications of such a change. See the same Table under Ninth month 24, 25, and consult the fore part of the following Table. See also Tab. 147, Eighth month 27, and Ninth month 1, 4, 15, with the Notes. A change towards dryness *during rain or snow* is favourable, at whatever time it occurs. See Tab. 107, Sixth month 14, the Note; and 126, Eleventh month 10, with the Note.

But the most valuable prognostics are afforded by a progress *from day to day,* towards the moist extreme of the season. Numerous instances of this gradation occur in the course of the Tables: and it is observable, that a retrograde movement towards dryness often takes place during the wet or showery weather, which the preceding advance towards the moist extreme had prepared us to expect. It will suffice to bring the following cases in proof. Tab. 137, Eleventh month 27, to 138, Twelfth month 19; and Tab. 140, Second month 13, to the end.

Such are the results of the few observations which I have incidentally made on the movements of this instrument, in connexion with other indications: and they tend to shew that a regular attention to it in this way would reward the pains of the observer. It would be necessary to complete success, to ascertain the mean proportion of the scale through which the index *recedes,* in a given time after sunrise in each season; by comparing the daily quantity with which, a judgment might often be formed as to the proximity of rain. The more palpable sign of this, given by the advance towards moisture in the forenoon, would then likewise be frequently found useful.

OF THE DEW- OR VAPOUR-POINT.

I had given so little practical attention to this part of the subject of Hygrometry, that it would have been suffered to pass unnoticed here, but for the appearance of a new instrument, adapted to the more ready discovery of the vapour-point; and which has come under my notice only since the last section was finished at press.

The introduction of this process into Meteorology is due, in common with many other original ideas and operations, to my friend John Dalton. It consists essentially in ascertaining, by means of a cold body, the temperature at which the vapour diffused in the atmosphere will begin to be decomposed and to deposit its water. The familiar fact of the dew which, in certain seasons, immediately forms on the outside of a glass vessel, newly filled with water from a deep well, may serve to illustrate the process: the intention is, to discover the precise temperature at which a body will begin thus to elicit water from the air; and by comparing this with the temperature of the air, to judge of its approach towards saturation: it being manifest, from our knowledge of the superior attraction of the permanent gases for heat, that the moment the air itself shall arrive at the temperature indicated by the cold body, it will perform the same office—it will rob the vapour diffused in it of a portion of its constituent heat, and separate the water in *dew, mist,* or *rain.*

Only three experiments of this kind are recorded in my observations in the First volume. In the first,

z

Tab. 1, I found the dew-point but one degree below
the temperature of the air at noon. On that day
there was no sensible evaporation by the guage, and
the rains, which had prevailed some time, continued
for several days after. On the second occasion, Tab.
67, I found it at 2 h. 30 m. p. m. within 3° of the
temperature of the air. " In an hour afterwards, it
began to rain steadily, and there fell more than half
an inch in depth." The third time, Tab. 83, " the
air was so loaded with vapour at 9 p. m. as to deposit
water on a glass vessel cooled to 58°, the lowest temperature of the following night being 53°. At this
this time it began to rain heavily, ceasing at 10, with
thunder and lightning still in the North." In effect,
during this and the four following days, there fell
above 2 inches of rain.

The time and trouble required to perform the operation on purpose (for an accidental deposition on a
glass has sometimes suggested the comparison of the
temperatures) have prevented frequent experiments
on my part. It is therefore with some satisfaction
that I notice the introduction of *Daniell's Hygrometer,*
of the construction and uses of which an ample account is given in the Quarterly Journal of the Royal
Institution, No. 16, together with a Meteorological
Journal of four months, containing its indications,
by the inventor.

In this instrument, the design of which appears to
have been taken from the model of Leslie's Hygrometer and Wollaston's Cryophorus, the requisite degree of cold is produced at all times, with ease and
certainty, by means of the evaporation of a few grains
of ether. The ether being dropped on the surface
of a hollow sphere of glass, covered with muslin, and
full of the vapour of ether, an immediate condensa-.

tion of the latter generates a vacuum within it : into this a second evaporation of ether instantly takes place from a second naked bulb partly filled with that liquid, and connected with the former by a tube. The cold thus produced in the second bulb, causes in a short time a visible deposition of dew from the air on its surface : the temperature at which this effect begins, is indicated by a small included thermometer ; and a second thermometer, for the temperature of the atmosphere at the moment, is attached to the pillar which supports the bulbs and tube.

It is obvious that by such an instrument greater facility is afforded to the Meteorologist, of satisfying himself respecting the state of the vapour constantly diffused in the atmosphere, and of drawing from this, *in conjunction with other evidence,* a more certain prognostic, in critical seasons, of wet and dry weather. It is not likely, any more than the Barometer, always to answer this purpose when used *alone,* there being evidently other conditions necessary to the production of rain at a given station, besides the present saturation of the air immediately incumbent on it.

The constant, though minute, expenditure of ether for the experiment may prove to some observers an objection to its use. This may be in part obviated by using a bottle fitted with a tube ending in a capillary opening, and closed in the middle, or near the fine part of its bore, with a stop-cock, which would prevent any unnecessary waste of this volatile fluid ; as the heat of the hand would suffice to expel a sufficient quantity upon the ball. It is also possible that some future improvement of the instrument may enable us to produce 20° of cold, which seems to be all that is wanted in the driest season, without the waste of any material at all. To those who may be occasionally engaged in experiments to find the force of vapour and quantity of moisture present in different seasons and places, and at different elevations in the atmosphere, I have no doubt that this elegant little instrument, which is portable (inclosed in a mahogany box) in the pocket, will be a valuable acquisition.

OF THE RAIN.

The position or elevation of the guage affects the product nearly as much, in the case of Rain, as in that of Evaporation, but in a different way. The Evaporation is increased, as has been shewn, by elevating the guage, but the product of Rain in this case is usually diminished: insomuch that when the guage is transferred from the ground to the house-top, the average falls off by about a fourth part.

I have treated this subject under Table 64 of my First volume; where are detailed the results of experiments made during twenty successive days of wet weather, in the autumn of 1811, wirh a view especially to discover a rule for correcting former results, as also to ascertain the circumstances under which these differences take place. I have since seen no sufficient cause to abandon the conclusion I then came to; that, when rain takes place with a turbid atmosphere, a considerable and variable proportion of the water is actually separated from the vaporous medium, at a height not exceeding 50 feet (or that at which my upper guage was fixed, which was 43 feet) and that this portion, consequently *must* be deficient in the upper guage. But in showers from an elevated region, falling through an air which is not itself undergoing decomposition, the products ought to be (as is the case in some instances) alike in both guages.

In the reasoning connected with those experiments, I did not advert to a possible constant effect of the wind, in lessening the product of the guage in this

more exposed situation; but contented myself with
proving by experiment "that rain may be drifted as
well as snow;" or that a portion of the general receiving surface may be robbed of a part of its rain,
by deflected currents in the moving atmosphere, which
transfer it to another place; which it was shewn happens especially where two guages are placed, the one
on the windward the other on the leeward parapet
of a building; the latter being redundant, while the
former is deficient in product, the level considered.
What proportion of the deficiency in a guage placed
on a building should be ascribed to this cause, it may
be difficult in all cases, without experiments carefully
made on the spot, to decide. It would undoubtedly
vary, according to the position of the guage, with
respect to the wind by which the rain might be carried
at the time. But the question of this difference has
lately been discussed in the Journals, on principles
purely mathematical. By some, and among them a
meteorologist of note in France, *Flaugergues*, it has
been attributed wholly to the effect of the wind, in
giving an oblique direction to the streams of rain;
in consequence of which, it is contended, the funnel
or mouth of the guage, actually presents a smaller
aperture, in proportion as the rain comes more
obliquely: just as if we were to incline the funnel to
one side under a rain falling vertically, in which case
it is manifest that less and less rain would enter as it
became more inclined, until in a perfectly horizontal
position, the whole would pass by to the ground.
But in reply to this, it is said, and I think very justly,
by *Meickle*, (in Thomson's Annals for October 1819)
that in the case of rain falling with a wind, "the
horizontal distance of the lines in which the rain falls
is absolutely independent of their inclination, being

accurately the same, where the wind runs steadily 60 miles an hour, as if it were a perfect calm." " In strictness (this writer further observes) *the drops fall in curves,*" but supposing them to pursue a right line " it is plain, that a guage of the width shown at *a b* will there receive the drops, falling obliquely, just the same as after they become perpendicular in the calm, at *c d.*" And it is equally obvious, from a comparison of the space *b–e* with the space *a–b,* that *of rain so deflected,* a guage with its aperture inclined in a suitable direction, would receive *much more* than the quantity then actually falling on the general surface : consequently that, *with reference to this stand- ard,* the aperture in the direction *a–b* would receive too little.

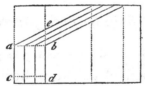

It appears therefore that no allowance is required, in any position of the guage as to height, for the simple obliquity of the whole body of a shower ; the rain received by the entire horizontal surface included in its area (and of which the mouth of the funnel may be considered as a definite part) being the same whe- the descent be oblique or vertical.

Mean Annual depth of Rain about London.

The Mean annual depth of Rain in our district is very nearly 25 inches. This being the largest average that has yet, so far as I know, been assigned to Lon- don, it will be proper to shew the ground on which it rests.

On the evidence of the experiments in the autumn of 1811 already alluded to, I assume for the present, that the rain on the ground is to that collected at 43 feet above it, as 37 to 28 : but in what proportion the difference may increase with a greater elevation, I have not the means of deciding : it is probable that the deficiency at greater elevations would be found in a rapidly diminishing proportion to the height. The guage of the Royal Society is stated to be 75 feet 6 inches above the surrounding ground. But so much of this surrounding ground is covered with buildings, that we may consider the *difference* between the two elevations as nearly done away in regard to practical effect, and I shall therefore neglect it

Annual Rain for twenty-three Years.

The following then is a statement of the annual Rain by the guage of the Royal Society for ten years previous to my own account. In a second column I apply to these amounts the proportion of difference above assumed, to bring out *an estimated result at the surface of the earth.*

Year		Rain at 75 feet.		Rain at the ground by estimate.
1797	....	22·697 in.	....	29·996 in.
1798	....	19·411	....	25·650
1799	....	19·662	. .	25·982
1800	....	18·925	....	25·008
1801	...	19·197	....	25·367
1802	...	13·946	....	18·428
1803	...	17·922	....	23·682
1804	....	20·973	....	27·714
1805	....	20·396	....	26·951
1806	....	20·427	....	26·992
	Averages	19·355 in.		25·577 in.

I find the average of seventeen years rain (with some interruption in the series of years,) from 1774 to 1796, to be, by the same authority, 19·762 inches; none of the products composing it appearing to have been collected at the ground.

The *Monthly* results of my own Register for the ten years from 1807 to 1816, are detailed in the General Table D. Many of these, in the fore part of the series, are marked as having been obtained at 43 feet elevation: to the annual totals of these results the same mode of correction has been applied, so far as it was required, and they come out as follows: the Mean Temperature of each year is annexed.

Year		Rain at the ground.		Mean Temp.
1807	....	18·01 in.	..	48·367°
1808	....	23·52	....	48·633
1809	....	24·18	..	49·546
1810	..	27·51	....	49·507
1811	....	24·64	....	51·190
1812	....	27·24	....	47·743
1813	....	23·56	.. .	49·762
1814	....	26·07	....	46·967
1815	....	21·20	....	49·630
1816	...	32·37	....	46·572

Average 24·83 in.

Lastly, the results obtained at or near the surface, for the years since elapsed, run thus:

1817	....	24·80	....	47·834
1818	..	25·95	.	50·028
1819	....	24·30	....	50·116

Average 24·87

The General average of all the results thus obtained for the ground, comprehending a period of twenty-three years, is 25·179 inches: I shall apply

2 A

presently to the several years a definitive correction, through the medium of the monthly results, which brings out an average still nearer to my own above stated.

Wet and Dry Years.

The greatest depth of rain in 23 years fell in 1816. Next to this, for wetness, appears the year 1797.

The driest year in this period was 1807, and next to it, 1802.

About one *year* in five may be said to be subject to the dry extreme, and one in ten to the wet.

After an extreme wet year, in 1797, we meet with four years in succession with an amount of rain very near the average of the climate, and then an extreme dry year: and since the extreme wet year of 1816, we have again had three years in succession near the average.

Connexion of the Annual Rain with the Mean Temperature.

In the series of years from 1810 to 1816, the reader will find, on comparing the rain with the mean temperature, that the warm years were uniformly dry, or below the average in rain, and the cold ones uniformly wet, or above the average. This is a very natural coincidence; but do the effects depend on the alternate warmth and coldness of those years, or rather on the mean temperature considered absolutely? In reference to this question, I may remark, that the mean temperature of 1797 was 49·398, and that of 1816, 49·433 (Royal Society), and their respective amounts of rain at the ground as above stated

29·996 and 32·87 inches. Again, as to dry years,
1802 had a mean temperature of 50·200° with 18·428
inches of rain, and 1807 a mean temperature of
50·733°, with 18·01 inches of rain. There is therefore probably a close connexion between the mean
temperature of many years, and the rain at the earth's
surface which attends them.

*Average proportions of Rain in each month of the year:
definitive correction of the amounts for the height.*

The product of rain for the same month in different
years varies, in each of the months, greatly. This is
a fact to which common observation is perfectly competent; and it is scarcely necessary to refer the reader
to the General Table D for the proof in figures. Having constructed a similar Table from the Monthly
results in the Philosophical Transactions, from 1797
to 1806, I found the same variety in them also. The
same month which in one year affords 5 inches of rain,
in another exhibits not a quarter of an inch; or even
(as appears in two instances) none at all, the few drops
that fell having been inappreciable by measure. It
became a question therefore, as before in the case of
temperature, what should be regarded as the mean
quantity; or the standard of comparison to which the
product of a wet or a dry month should be referred.

The following are the *averages* of rain for the
respective months as obtained by actual observation:
the first series on a period of ten years, from 1797 to
1806, by the guage of the Royal Society; the second
on a similar period, from 1807 to 1816, from the results in my own register. In a third column are
inserted the *average number of days on which any rain
fell,* in each month of the latter period.

Average Rain for the Month by observation at different levels.		Number of days on which it rained
	1797—1806	1807—1816	
1. Jan.	1·341	1·633	14,4
2. Feb.	0·911	1·486	15,8
3. Mar.	0·755	1·422	12,7
4. April	1·282	1·550	14,0
5. May	1·340	1·921	15,8
6. June	1·708	1·928	11,8
7. July	2·555	2·578	16,1
8. Aug.	1·925	2·102	16,3
9. Sept.	1·833	1·522	12,3
10. Oct.	1·671	2·740	16,2
11. Nov.	2·400	2·407	15,0
12. Dec.	1·631	2·093	17,7
Totals	19·352	23·382 in.	

I have not introduced here any correction for the difference of level, because it will be more interesting first to compare the quantities as found.

In general, the average at the lower level exceeds; but in two cases the higher equals, and in one it exceeds the lower: which considering they are for different periods of years was to be expected.

In the months of September and November this is clearly due to excessive rains in those months, in four out of the ten years of the first series.

In October, the lower level is disproportionately in excess, from the same cause operating in the latter series.

Setting aside these cases, let us advert to the Seventh month. Here the rains are alike in both averages: and on examination, the respective Tables of results furnish no adequate reason for this. In 1806 there fell indeed in this month 4·889 inches at the higher level, but in 1800 the month was absolutely dry: the two taken together make an average near the one

in the Table; and in other respects the results in each
series present a very similar range of quantities.

If we recede again from this to the contiguous month,
in either direction, we find in one of these a deficiency
of an eighth, in the other of a ninth, in the higher
level: which deficiency in the months of Winter and
Spring becomes more considerable, the proportion in
one instance amounting almost to the half of the sum
of the lower product.

The probability is therefore very strong, on the
evidence afforded by these averages taken in conjunction with my experiments in 1811, that the deficiency
in the rain collected at the higher level, from whatever cause or causes proceeding, is very small in the
midst of summer, and increases as we recede in either
direction towards winter. In the former season, the
showers fall mostly from elevated clouds, and the
lower atmosphere is generally clear of that misty precipitation which, in the winter months, must contribute something considerable towards the product at
the ground. Add to this that the effect of strong
winds, in whatever way it robs the higher guage,
must be far most considerable in the latter season.

On these considerations I have ventured to construct a Table of Monthly amounts of Rain, *corrected
for the surface of the ground*; in which the rate of
allowance is made to increase from 0·05 on an inch
in the Seventh month, to 0·50, in the First, and decrease again through the remaining months in like
proportion; the rate of gradation being 0·10 in Spring
and Autumn, and 0·05 in Summer and Winter. These
form the General Table H, at the end of the volume.
I am aware that many of the cases, taken singly,
cannot be accurate as they stand: there being seasons
in which our summer rains resemble the storms of

winter, and others in which the latter season has summerlike showers: but the whole twenty years from 1797 to 1816, thus modified, afford an average of 24·808 inches per annum ; which, it will be seen, differs by but a very small fraction from the averages found at the surface, or corrected on the evidence of experiment.

The annual results arising out of this mode of correction differ somewhat in amount from those obtained by estimating the deficiency on the whole year. The reader may give the preference (if he pleases) to the latter, or corrected results, without its materially affecting the consequences I have drawn from the *estimated* ones.

It might have been expected that I should have here carried on the parallel between the results of the Royal Society and those of my own guage, through the remaining years published in the Transactions, and thus have settled the difference on the basis of actual observation in each case : but it is with regret that I acknowledge myself defeated in this object by an apparent falling off, of late years, in the conduct of that Register.

The years 1807, 1808, and 1809, present indeed an average of rain which agrees sufficiently with the former averages, and is proportionate to the estimated results at the ground : but in 1810 we have no account of the rain at all; and in 1811, for the latter half of the year only. From 1812 to 1818, the annual average sinks at once to about 15 inches, the former averages deduced from long periods of years, having been about $19\frac{1}{2}$ inches! We have however a statement of the annual rain in 1812, 1813, and 1814, " by another Rain-guage, placed a few feet distant from the former, and 11 feet 6 inches lower," the average of which is 20·349 inches: but the monthly results of this guage have been neglected in the body of the Register, except in some instances where they appear to have crept in by inadvertence, or to be stated on the opposite page by way of contrast to the higher ones, though without notice of this circumstance to the reader. The only reason which I can assign to myself for this extraordinary deficiency in the higher guage is, that the rain being now measured (as it seems) only at long intervals, about a fourth part of that which is actually collected in the year escapes by evaporation.

If this learned and highly respectable body feels the subject of the weather no longer worthy its notice, would it not be better at once to dismiss the Register from its Transactions? But if, as in some sort the representative of our country in matters of science, it should be disposed to entertain an honourable emulation on this point with the Royal Observatory at Paris,* it will be necessary that much greater attention be paid than for several years past, both to the providing the requisite instruments and the due attendance upon them. For it is not in the article of rain alone, that defect or inaccuracy has introduced itself, to the degree almost of suspending confidence : an imputation which after being thus obliged to support (it having been already publicly advanced†) I should be equally ready to contribute in any degree to do away.

---

* The " Meteorological Observations made at the Royal Observatory of Paris" are quite a model in point of care and exactness. They are published monthly in the " Journal de Physique ;" and the results at least, ought to be given, in our own language and measures, in the periodical Journals of this country.

† By Dalton, *Manchester Memoirs*, vol. 3, second series, p. 490.

To return to the subject of the proportion of rain which falls in the different seasons—the following Table exhibits the monthly averages for the level of the ground for two periods of ten years each; the first set, corrected from observations at 75 feet elevation, the second in part corrected from observations at 43 feet, but chiefly as obtained at the ground. The third column exhibits the two averages incorporated.

Monthly averages of Rain, corrected for the elevation.

Month	1797—1806	1807—1816	1797—1816
1. Jan.	2·011	1·907	1·959 iu.
2. Feb.	1·320	1·643	1·482
3. Mar.	1·057	1 542	1·299
4. Apr.	1·666	1·719	1·692
5. May	1·608	2·036	1·822
6. June	1·876	1·964	1·920
7. July	2·683	2·592	2·637
8. Aug.	2·117	2·134	2·125
9. Sept.	2·199	1·644	1·921
10. Oct.	2·173	2·872	2·522
11. Nov.	3·360	2·637	2·998
12. Dec.	2·365	2·489	2·427
Totals ...	24·435	25·179	24·804

The wettest month, in a long run of years, appears by this method to be the Eleventh, or November; but on the evidence of the latter period, which has the most of actual observation for its support, the Tenth may dispute the precedency in this respect. Yet in perusing the amounts in the third column we see the rain falling off in nearly equal proportions in each direction for two months from the Eleventh; which, as there was not the smallest adjustment to produce this effect, may seem to prove the mode of correction employed nearly accurate.

The next amount of rain is the Seventh, or July

From this month we have a diminishing series of amounts to the Third, or March, which is the driest; having only half as much rain as the Seventh, and a little more than two-fifths of the quantity of the Eleventh month. In the first series of years, the Third month is comparatively dry, and the Ninth wet: in the second series the Third is wet, and the Ninth dry, compared with the general average: the reader will find many examples of the same contrast in the particular years on which the averages are founded.

From the Third month, proceeding forward, we see the rain grow larger in amount to the Seventh, then less to the Ninth, and larger to the Eleventh again.

The following diagram exhibits the gradation of Rain through the year, on a scale of half the depth.

Fig. 11.

W—N, S—W, N—E, N—E, S—W, W—N, W—N, W—N, S—W, S—W, S—W, S—W.

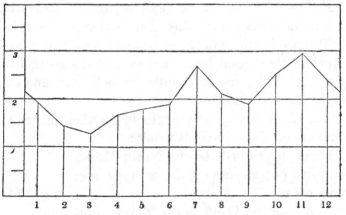

Proportions of Rain in the different seasons.

In an Essay read in 1818, by my friend John Dalton, before the Literary and Philosophical Society of Manchester, the author concludes from different averages, " that the first six months of the year must be considered as dry months, and the last six as wet months; that April is the driest month in the year, and the sixth after, or October, the wettest."

My own averages perfectly coincide with his first proposition, if only the slight difference be neglected by which September falls below the mean.

If in the results here advanced the driest average falls a month earlier, and the wettest a month later than in his statement, it may perhaps be attributed to the difference, in exposure and latitude, of the two stations, Manchester and London. The latter may be thought to lose the hibernal rains earlier, from its more forward spring, and to receive them more slowly in consequence of a more prolonged summer.

With regard to the proportions of rain in the former and latter half of the year, they stand thus by the average in the third column :

For the first six months (Jan.—June)   10·174 in.

For the latter six months (July—Dec.)   14·630

The two portions of the year thus divided are nearly equal in their total heat : the sum of the mean temperatures of the first six months being 280,32, and that of the last six 315,47.

But if we divide the circle in another place, we shall have a very different result :

From the Fourth to the Ninth Month inclusive (April—Sept.) the average rain amounts to   .   . . .   . .   12·117 in.

And from the Tenth to the Third inclusive (Oct.—March) to .   . .   12·687

Now the sum of the mean temperatures of the first six months in this series (or the summer half year) is 345,94;

and that of the remaining six months (or the winter half-year)   .   .   .   .   ,   .   .   ,   .   .   249,85.

Thus in dividing the year in one way, we have very
unequal amounts of rain for the two moieties, with
nearly equal amounts of heat; and in another way,
very unequal temperatures, with nearly equal amounts
of rain. If instead of taking the results of whole
months, the division had been made at the days of
extreme and mean temperature, the contrast would
probably have been still more perfect, at least as to
equal temperatures with unequal rain: but the more
direct method suffices for the object.

The solution of the whole case seems to be as follows:

In the former half of the year (that is, from some
time in the First month,) the mean diurnal heat is
advancing; or if it be kept down by a succession of
Northerly winds, these arrive in a state to promote
evaporation, and dry up, rather than deposit moisture: in the latter half (that is, from some time in the
Seventh month) the heat is declining; or if it be sustained towards the close of the year, by Southerly
winds, these coming into a colder latitude deposit
their water in consequence. Thus there is prevalent,
during the former half of the year, a cause which
powerfully counteracts the production of rain; and
during the latter half, a cause which more than any
other promotes it: the quantities of effect are therefore very unequal.

But in dividing the year at the points of mean temperature, we set these causes in opposition to each
other, in either moiety. The effect of the depression
of the mean temperature in the last three months of
the year, is counteracted by an elevation in the first
three months; and the elevation continued through
the three months of spring, balances the depression
which ensues in the course of the three months after

Midsummer. The quantities of effect in the two moieties are therefore equal.

Consistently with this statement are the proportions of rain for the four quarters of the year, taking (as before in the case of Evaporation) the Twelfth, First, and Second as *Winter*, and the remaining months in classes of three, for the other seasons:

	Rain.	Mean Temp.
Winter	5·868 in.	37·20°
Spring.... ...	4·813	48·06
Summer	6·682	60·80
Autumn	7·441	49·13

The rain, it appears, is not any more than the evaporation, proportionate to the *mean Temperature of the season*. Yet if we add together the mean Rain and mean Evaporation for each season, the *sum of the two* will be found in pretty near proportion to its mean Temperature: the sums are,

For Winter	9·455
For Spring..............	13·669
For Summer	18·262
For Autumn	13·885

Proportions of Rain by day and night, &c.

In the early part of my observations I frequently measured off the rain which fell by day, and separated the result from that of the following night. As this was done only through the years 1807 and 1808, which are both below the average in rain, and with the higher guage only, I do not attach much importance to the results.

Of 45 inches of rain, which fell in the space of 31 Lunar revolutions, I divided 21·94, and found 8·67 to have fallen with the Sun above the horizon, and 13·27 during his absence. According to this experiment

the rain by day makes only two thirds of the quantity
that falls by night.

The greatest rain in twenty-four hours that has
fallen under my observation (or rather that of my
assistant, for I was shut up in the pacquet going to
Helvoetsluys,) was on the day and night of the 26th
of Sixth Month 1816. On this occasion, the night in
particular was very wet and stormy : the guage at our
Laboratory, Stratford, near London, collected 2·05
inches, between 9 a. m. the 26th and the same hour
the following morning. I have already noticed the
character of that season, which was at once the coldest and the wettest in 20 years. See Table 120, &c.
in the First Volume of this work.

In reverting to the column, p. 188, entitled " Number of days on which it rained" in each month, the
reader will perhaps be struck with the fact, that in
our climate, on an average of years, *it rains nearly
every other day, more or less.* He will perceive however, that the number of days (of twenty-four hours)
on which there falls any rain, is less in the longest
days than in the shortest, in the proportion of two to
three.

The propensity to frequent dripping, even in this
dry corner of our island, I consider to be connected
with our moderate and variable temperature. In climates the mean temperature of which, from the circumstance of latitude alone, departs farther in either
direction from the mean Temperature of the Earth, it
is probable the number of *dry* days will be found
greater, in proportion as the climate is hotter or
colder than our own. It will be an interesting enquiry
for those who are sufficiently zealous in these pursuits,
and who possess the requisite leisure, to follow out
the comparison through the many registers of the

weather already published. The circumstance of elevation above the sea, which when not excessive greatly tends to promote rain, will require the exclusion of some registers from this comparison; unless indeed the mean rain for a certain range of latitude be deduced from the whole of the registers kept at different elevations within its limits.

Popular adage of " Forty days rain after St. Swithin" how far founded in fact.

The opinion of the people on subjects connected with Natural history is commonly founded in some degree on fact or experience: though in this case, vague and inconsistent conclusions are too frequently drawn from real premises. I have already stated under Tab. 83, that the notion commonly entertained on this subject, if put strictly to the test of experience *at any one station* in this part of the island, will be found fallacious. To do justice to popular observation, I may now state that in a majority of our summers, a showery period, which with some latitude as to time and local circumstances may be admitted to constitute daily rain for forty days, does come on about the time indicated by this tradition: not that any long space before is often so dry as to mark distinctly its commencement.

The tradition, it seems, took origin from the following circumstance. Swithin or *Swithum*, Bishop of *Winchester*, who died in 868, " desired that he might be buried in the open church-yard, and not in the chancel of the minster, as was usual with other bishops, and his request was complied with: but the monks on his being canonized, considering it dis-

graceful for the saint to lie in a public cemetery, resolved to remove his body into the choir; which was
to have been done with solemn procession, on the
15th of July: it rained however so violently for forty
days together at this season, that the design was abandoned."*   Now, without entering into the case of the
Bishop, (who was probably a man of sense, and
wished to set the example of a more wholesome, as
well as a more humble mode of resigning the perish-
able clay to the destructive elements) I may observe
that the fact of the hindrance of the ceremony by the
cause related is sufficiently authenticated by tradition:
and the tradition is so far valuable, as it proves that
the summers in this Southern part of our island, were
subject a thousand years ago to occasional heavy rains,
in the same way as at present.   Let us see how, in
point of fact, the matter now stands.

In 1807 it *rained* with us on the day in question,
and a dry time followed.   In 1808, it again rained on
this day, though but a few drops: there was much
lightning in the W at night, yet it was nearly dry to
the close of the Lunar period (at the New Moon) on
the 22d of this month, the whole period having yielded
only a quarter of an inch of rain: but the next moon
was very wet, and there fell 5·10 inches of rain.

In 1818 and 1819, it was *dry* on the 15th, and a
very dry time in each case followed.   The remainder
of the summers occurring betwixt 1807 and 1819, appear to come under the general proposition already
advanced: but it must be observed that in 1816, the
wettest *year* of the series, the Solstitial abundance of
rain belongs to the Lunar period *ending* (with the
Moon's approach to the Third quarter) on the 16th of

* Times Telescope, or Guide to the Almanac. 1814.

the Seventh month; in which period there fell 5·13 inches, while the ensuing period, which falls wholly within the forty days, though it had rain on twenty-five out of thirty days, gave only 2·41 inches.

I have paid no regard to the change effected in the relative position of this so much noted day by the reformation of the Calendar, because common observation is now directed to the day as we find it in the Almanac; nor would this piece of accuracy, without greater certainty as to a definite commencement of this showery period in former times, have helped us to more conclusive reasoning on the subject.

Solstitial and Equinoctial Rains.

Our year then in respect of *quantities* of rain, exhibits a *dry* and a *wet* moiety. The latter again divides itself into two periods distinctly marked, as the reader will perceive by viewing the two elevations of the curve in fig. 11, p. 193. The first period is that which connects itself with the popular opinion we have been discussing. It may be said on the whole to set in with the decline of the diurnal mean temperature, the maximum of which, we may recollect, has been shewn to follow the Summer Solstice at such an interval, as to fall between the 12th and 25th of the month called July. Now the 15th of that month (or Swithin's day) in the old stile, corresponds to the 26th in the new: so that common observation has long since settled the limits of the effect, without being sensible of its real cause. The operation of this cause being continued usually through great part of the Eighth month, the rain of this month exceeds the mean, by about as much as that of the Ninth falls below it.

The latter wet period corresponds to the second great elevation in the curve. It begins by a large addition to the amount of rain in the Tenth month ; its middle and wettest part falls in the Eleventh ; and it goes off by a similar gradation of amount in the Twelfth. As the former period apparently takes its origin from the Summer Solstice,* (though, like the highest temperature, not developed till after a certain interval,) so this occupies much of the interval between the Autumnal Equinox and the Winter Solstice, its termination being apparently fixed by the latter. I propose therefore to distinguish the two by the terms, Summer or *Estival,* and *Autumnal* Rains; meaning thereby not the entire quantities of rain falling in the midst of two seasons otherwise dry, which would be inapplicable to our climate, but simply the *excess* of rain, which on a mean of years the two periods afford us.

---

* The great rains in Abyssinia and the neighbouring countries, on which depend the annual overflowings of the Nile, have a period nearly coincident with that of our own *Estival* rains. The dry season, according to Bruce, gives place to light rains about the Vernal Equinox, but there falls no considerable quantity in any place until the Sun arrives at the zenith: from which time, until in returning southward he becomes again vertical to them, all these parts are subject to heavy rains; which cease as to the whole country a little before the Autumnal Equinox. The following are the Monthly proportions of the rainy months on a medium of two years, according to a Journal kept by Bruce's assistant *Balugani,* at Gondar in 1770, and at Koscam in 1771.

May	2·609 in.	Aug.	12·794 in.
June	5·347	Sept.	5·086
July	12·224		

And it is remarkable that in these parts of the world they have a second rainy period, of much less extent, and as it seems, less certain than the former; which falls about the close of October and beginning of November. Thus in Summer *we* partake in a less degree, and with

The *Equinoxes* themselves are in our climate comparatively *dry,* the Vernal especially; and they are

some uncertainty, of the operation of the causes which produce the
Tropical rains; and in Autumn, the Tropical regions are affected, in
a lower degree, by the causes of our more complete precipitation.

I attach little to the shew of minute accuracy with which this Journal of Bruce's is got up; there being, after all, some inconsistencies
apparent, owing perhaps to the press: but I think there is internal
evidence of its being a real register, which may be depended on for
general results. His account of the manner of the falling of those
Abyssinian Thunder-showers, in a country so elevated that the Barometer stood at 22 inches is, though very unphilosophical, an interesting
record: since the phenomena which he there attempts to describe, may
often be witnessed day after day in our own climate, though so near
the level of the sea, at the same season. " Every morning in Abyssinia is clear, and the sun shines. About nine, a small cloud not above
four feet broad, appears in the East, whirling violently round as if
upon an axis; but arrived near the zenith, it first abates its motion,
then loses its form, and extends itself greatly, and seems to call up
vapours from all opposite quarters. These clouds, having attained
nearly the same height, rush against each other with great violence;
the air impelled before the heaviest mass or swiftest mover, makes an
impression of its own form in the clouds opposite, and the moment it
has taken possession of the space made to receive it, the most violent
thunder possible to be conceived instantly follows, with rain. After
some hours, the sky clears with a wind at North; and it is always
disagreeably cold when the Thermometer is below 63'. When the
sun is in the Southern tropic, 36 degrees distant from the zenith of
Gondar, it is seldom lower than 72°, but it falls to 59° when the sun
is immediately vertical." Bruce's Travels, Book 6, chap. 15 and 19.

The whirling cloud, which makes such a figure here as the precursor
of the storm, I believe to be the first appearance of the *Cumuli,* which
it is afterwards made to call up around it, and which he may have
observed at times coming on from the windward with a rapid intro-
version of the *apex* upon the *body* of the cloud, as sometimes happens
here before heavy rain. The rest is merely a confused description of
the gathering of a thunder-storm; in which, instead of the air carrying
the clouds, the latter are made, like projected solids, to propel the
air before them, and thus remove the neighbouring clouds in order to
take their places!

attended with the remarkable circumstance of an occasional anticipation (as it were) by the *Vernal,* of a share of the rain which might be expected to accompany the *Autumnal* Equinox.   Were it not for this curious connexion, the Ninth month would not be *dry,* but would have its rain above the mean of the year.   The reader will find this translation to have happened in the years 1801, 1804, 1810, 1812, 1814, and 1815; by comparing in Tables H and D, the amount of rain for the Third and Ninth months respectively.

Connexion of the Rain with the Winds.

I have placed at the head of the diagram, p. 193, the prevailing wind for each month, or rather the *class* which exhibits the highest number for the month, in the average on ten years given in the Table, p. 158. It appears at once that a wind between North and East is connected with our driest season, about the Vernal Equinox ; and a wind between South and West with the wet season following the Autumnal.

There is a regularity in the succession of the winds in the first six months, of which till I came to this part of the work I had not suspected the existence. The classes run thus, W—N, S—W, N—E, N—E, S—W, W—N.   After this, the class W—N prevails during the Summer, and the class S—W through the latter four months of the year.

The short notice respecting the temperature proves that a warm vaporous current, probably from the neighbouring Arabian gulf, is decomposed by the action of a colder Northerly one ; and the whole of the circumstances represent, on a grander scale, the weather of a wet thundery summer in this climate.

The connexion of a different class of winds with the *Autumnal* from that which prevails during the *Estival* rains, may be admitted as a proof, that the two periods which have been described, are really distinct effects, produced by different arrangements of the causes of rain in the atmosphere.

On summing up the horizontal columns of observations on the wind in Table D, which comprehend a space of ten years, I found the following to be the annual amounts of the several classes; which are here put in comparison with the *corrected amounts of rain* for those years.

	N—E	E—S	S—W	W—N	Var.	Rain.
1807	61	34	113	114	43	20·14 in.
1808	82	38	108	103	35	23·24
1809	68	50	123	91	33	25·28
1810	81	72	78	83	41	28·07
1811	58	59	119	93	36	24·64
1812	82	66	93	91	34	27·24
1813	76	53	92	124	20	23·56
1814	96	65	91	96	17	26·07
1815	68	36	121	107	33	21·20
1816	64	66	106	102	28	32·37
Averages	74	54	105	100	32	25·18 in.

This Table affords some very striking results, as to the manner in which the several annual quantities are related to those of the Rain.

In the driest year of the whole, which is 1807, the class N–E has nearly double the number of the E–S; in 1815, the next for dryness, the same; and in 1808, which stands third, rather more than double.

In 1816, the wettest year, on the contrary, the class E–S *exceeds* the N–E: in 1814, it has two-thirds of the amount of the latter; in 1812, three-fourths; and in 1810, the remaining wet year, the amount comes within a ninth of the N–E, both classes being large, and the Westerly winds falling off in a remarkable manner to make room for them.

The year 1811, which has about an average of rain, has the features, in respect of winds, of a wet year. On examination, I find that 36 out of the 59 observations here forming the E–S class are put down as an E wind: and in 1809 and 1813, the two remaining years, both a little below average, the majority of the observations in the second class are of the same kind.

These proportions, then, confirm the relation, already exhibited in the diagram, of a NE wind to the dry weather; and they establish another relation between a SE wind and the rain, of our climate.

With regard to Westerly winds,—the class W–N, we may observe, falls off gradually during the three years following 1807, while the annual rain increases from year to year: and in four of the remaining years, its number is above the average in the dry years, and below it in the wet ones. There is therefore a manifest general relation of this class to our fair weather.

The winds between the South and West have no decided connexion with either a wet or a dry year.

This proposition may seem at variance with the connexion exhibited in the diagram, and with the remarks upon it, which appear at the beginning of this section : but the contradiction is apparent, not real, as will appear by what follows.

General ideas on the Rain of these Latitudes.

There are two ways in which we may conceive Rain to be produced in a temperate latitude: First, by the cooling of the whole mass of the atmosphere to a degree sufficient to decompose its vapour. This happens when, either the air flowing constantly from the South to the Northward, leaves the influence of the Sun behind it; or the Sun, declining in Autumn and retiring to the Southward, leaves the air to cool where it remains. In effect, both causes may be in action together, as is probably the case during some part of every autumn in these latitudes.

Secondly, by the cooling of a portion only of the air — from the intrusion, or the overflow of a warm vaporous current, from a lower latitude into our own; where it loses its heat, and has its vapour decomposed by our colder air.

In the first case, the rain will be formed in every part of the atmosphere, up to a certain height at least from the ground; where the vapour diffused through a rarefied medium can afford only a kind of hazy precipitation, which gradually descends upon the lower air. In the second case, showers, and hail with thunder, if the contact be very sudden between the currents, are generated; which fall from a greater height, and are commonly much less *continuous* than the other kind of rain.

Both of these modes of production, again, may be in action together for a time. A Southerly current, charged with vapour from a warmer region, may be passing Northward, at the same time that a Northerly current may be returning towards the South in its immediate neighbourhood: and these two may rase each other, the colder running in laterally under the warmer current, and causing it to flow over laterally in its turn, while each pursues in the main its original course. In this case the country, for a considerable space extending from about the line of their junction far into the Southerly current, may be the seat of extensive and continued rain.

In this case, if in any, we may admit the principle advanced by Dr. Hutton, that when two portions of saturated air mix together, the common *temperature* will be reduced to a medium between the temperatures of the two, while the *capacity* of the mixture, or its power of retaining water, will be much below the medium; and precipitation will ensue from this circumstance alone. This principle the Doctor thought applicable to every case of rain that could happen; and not only to these, but to the production of clouds and mists also, in whatever circumstances or situations they may be found.

It is certainly of great importance to establish general principles, on which we may reason conclusively respecting any case which may present itself in Nature; and when informed of the premises, be able to determine that such or such consequences *must* ensue: but I confess I doubt, notwithstanding the authority of more profound reasoners in its favour, whether Meteorology will really gain a step at present by adopting this system.

" That the quantity of vapour capable of entering into air increases in a greater ratio than the temperature," is a proposition which appears to rest on the basis of experiment: but " that whenever two volumes of air of different temperatures are mixed together, each being previously saturated with vapour, a precipitation of a portion of vapour (water) must ensue," is at present demonstrated by no experiment that I know of, and requires, I think, to be reconsidered. The reason given is, that the mean Temperature is not able to support the mean quantity of vapour ;* but are we sure that the Temperature in this case will be in the Arithmetical mean? We know that such is the result with homogeneous bodies, as with equal volumes of hot and cold water: but volumes of air saturated with water at different temperatures are in the case of heterogeneous bodies: they differ in composition, the warmer air containing the most of aqueous vapour: the specific heat of aqueous vapour is given at $1 \cdot 55$, while that of air is $1 \cdot 79$, water being unity. Such a mixture will therefore probably have a temperature differing from the Arithmetical mean; and *possibly* differing in such a way, as to prevent the precipitation of any water in consequence of the equal distribution of the heat in the mixture.

I am doubtful, secondly, of the fact of the intimate mixture of large masses of the atmosphere in the manner, and to the extent required by this hypothesis. The natural appearances are against it: we can often trace, during the approach and in the intervals of rain, the gradual descent or subsidence of a superiour current, which sooner or later manifests itself at the surface, by a wind there flowing in the same direction. Often, in summer, more than two of these may be detected; which, after quietly flowing over each other, without any extraordinary precipitation, or in some cases (as when observed by means of balloons) without any turbidness at all, come down in succession, during several subsequent days of fair weather. Supposing these to have been all saturated with water, what should have made them more liable to mix than in the case described? That they do not mix at all in the plane of contact, is not what I maintain; but

* Dalton, Manchester Memoirs, vol. iii. second series.

simply, that the intimate mixture of the whole elements of the atmosphere, from the height of some
thousand feet down to the earth, and this every time
that it rains largely, is not from appearances a probable supposition.

Many phenomena attending the production of dew,
mists and clouds, might be cited as adverse to the
opinion of such extensive intermixtures of the higher
and lower atmosphere: but leaving the question of
the *modus operandi* in the case, to be settled by those
who incline further to discuss it, I may state, as matter of experience, that the contact and opposition of
different currents *charged with aqueous vapour*, and
(by inference from their state as they manifest themselves in succession at the surface) *differing in temperature,* is largely concerned in the production of
our Vernal and Estival rains.

When after a suffocating heat with moisture, and
the gradual accumulation of Thunder-clouds, followed by discharges of Electricity, I observe a kind
of Icicles* falling from the clouds, then large hail,
and finally rain; and when after this I perceive a
cold Westerly or Northerly wind to prevail, I have a
right to infer, that the latter, aided by the electrical
energies, has been acting, *as a cold body in mass,* in
a sudden and decided manner, on the warm air in
which I was placed before the storm. Again, when
after a cold dry North-East wind I behold the sky
clouded, and feel the first drops of rain warm to the
sense; and after a copious shower perceive the air

* I witnessed such a phenomenon at Meriden near Coventry, on the
19th of Seventh Mo. 1803, when the large hail broke the windows in
that city. The reader will find a more striking instance of intense
sudden precipitation, in the account of the Gloucester Thunder-storm,
under Tab. 21, in my first volume.

below changed to a state of comparative warmth and softness, I may with equal reason conclude, that the Southerly wind has displaced the Northerly; manifesting itself first in the higher atmosphere, and losing some of its water by refrigeration in the course of the change. Doubtless *mixture*, in each case, obtains to a certain degree, and accelerates the effect; but it does not appear to me a *necessary* previous condition. On the contrary, the occurrence of rain, when the air is rather dry by the Hygrometer below (as sometimes happens) with the sudden increase of evaporation which often ensues upon rain, convince me that *temperature* may effect an occasional precipitation at the plane of contact, such as the general state of the atmosphere, had time been given for its operation, would have prevented. Nothing is more common than to see vapour issuing into a *dry* air (provided it be cold enough) decomposed by the contact of the latter, and yielding a copious *steam*, which is presently afterwards taken up again. In the same manner I suppose an occasional precipitation to take place, even to the degree of rain, by the mere circumstance of the sudden translation of a vaporous current into the midst of a cold medium; or of the irruption of the latter upon vaporous air at rest.

I make no use here of the effect of different electricities, which may obtain in currents brought from a distance and acting on those which they meet, to produce rain; because I am inclined rather to consider the electrical phenomena attendant on rain as secondary, and depending on the previous separation of watery particles in some degree of aggregation, by the great and universal cause of rain, *depression of the temperature of vapour.*

The previous reasoning is meant to apply to the

apparent anomaly, of a Northwest wind predominating in our wettest season in summer, and a Southwest
during the autumnal rains. I conclude from a careful review of the cases, that the former is not the
carrier, but the *condenser* of the vapour; which appears to be introduced at intervals only, from the
South and South-East. When the surplus vapour
has been disposed of in rain on these occasions, the
North-West resumes its sway, the atmosphere reco-
vers its transparency—et *claro* cernes sylvas *Aquilone*
moveri. Virgil, Georgic. 1 : but it is usually not long
before the returning clouds indicate the near approach
of a new supply of vapour—namque urget ab alto
Notus. Idem. The means by which we sometimes
escape the rains, and enjoy a dry and hot summer,
may be a subject of notice further on.

In the decline of the year the rain appears to origi,
nate, as before observed, in a somewhat different way.
The great body of the atmosphere is then usually
moving with some force from South-West to North-
East, while the Sun is declining to the Southward.
An air already turbid from beginning precipitation,
is further charged, below, by an excess of evaporation
from the agitation of much watery surface over which
it passes. Every calm interval then affords its shower,
followed by wind and evaporation again : and a succession of gales by night and cloudy days, charac-
terise the approach to the hibernal season. Exceptions however are found here, as in the former case ;
the autumnal rains being sometimes (though rarely)
scanty, and the weather more inclined to frost.

As however the whole of this reasoning has proceeded upon the evidence of general results and
averages, it may now be suitable to apply it to particular cases, of the connexion of rain with the vari-

ation and succession of the winds. The observations on the latter, having been conducted without the slightest previous intention of establishing such general principles as have been here advanced, the reader will be aware that many changes favorable to the hypothesis may have been overlooked. He will also be aware that a current from North or South may at times move through a considerable space in the higher atmosphere, and there be spent; manifesting itself below only by the precipitations which it occasions: also that the prevalent wind of the winter season, the South-West, must be allowed occasionally the same operation on an intruding vaporous current from the South-East, as has been ascribed to the more direct antagonist of the latter, the North-West.

Particular cases of Rain, &c. examined.

Tab. 1—2. In these two periods a mean quantity of rain was brought by strong SW winds. Once, when there fell 0·63 in. the nocturnal Temp. was depressed 12°, several hoar frosts ensued, and a positive electricity was manifested before the rain: all which indicates the interference by night of a Northerly wind.

A Thunder-storm (Dec. 2) seems to be connected with the ensuing dry winter.

Feb. 6—7 After a dry time of long continuance, the Fifth Month gave betwixt two and three inches in the first 15 days, *six* of which had variable winds. During these rains, which brought warm weather, (although from NE and E,) there was an abundant developement of electricity. On a sudden depression of Temp. at the close of the month, there fell 0·82 in.:

the wind getting to S and Var.: then, after two days fair with NW, the return of variable winds brought an inch and a quarter of rain. After which, with Westerly winds, the whole period, Tab. 8, was dry and *non-electric*; though upon the change from SW by W to NW, some thunder clouds made their appearance.

Tab. 10. In this period we have a rain of 0·93 in.: from the NW displacing the SE, attended with much lightning in the night: max. Temp. depressed from 81° to 72°.

Tab. 11. A rain of 0·65 in. clearly due to a depression of Temp. (by a Northerly current) of 10° on the day and night. And towards the close, immediately after the Autumnal Equinox, a wet week, apparently from a mere South wind, which *raised* the Temperature, while the Barometer fell.

Tab. 12. In the latter part of this period, we have seven wet days with a considerable depression of both Barometer and Thermometer, indicating an extensive decomposition of the aqueous atmosphere, the brunt of which appears by the note to have fallen upon the country about *Paris.* The winds are here a perfect mixture, viz. S, Var., NW, N, SE, E, N, SW, W, N: crossing the compass in both directions.

Tab. 13. Here we have 0·61 in. of *snow* by the NW supervening upon the SE: the latter wind is inserted upon the authority of the Philo. Trans. as my own observation is wanting. In the same period, 0·49 in. *snow* and *rain,* connected with 3 days *Var*

Tab. 15. The first days of 1808 presented an example of the effects of the interference of a partial current. A course of dry SW winds, of eighteen days duration, is interrupted in the middle by one day *South,* probably in its origin SE; the result is 0·65 in. of rain.

Tab. 18. After four days *variable* winds, 0·59 in. of water from *snow*, attended with positive and negative electricity. This was on the 20th of the Fourth Month, 1808. The 17th of the same month in 1807 was distinguished by a snow-storm, and the 19th by a succession of electrical Nimbi discharging dry hail: and the 20th, and 21st of the same in 1809, by plentiful snows, followed by hail and rain. In the two latter cases the NE and SE winds appear to have been in simultaneous action: in 1808 the winds are not specified. This analogy gives place to fair weather at the same season in 1810: but in the subsequent years, I find *hail* noted, for the most part, about the same time in the month. There is therefore probably a periodical current from the North at this season, in the higher atmosphere, the arrival of which is determined by the Sun's progress in North declination.

A case remarkably analogous, and which may prove important in regard to a future theory of the Atmospherical variations, occurs while I am occupied with the present sheet of this work. On the 19th instant (Second month 1820) after a considerable depression of Temperature for the season (there having been no snow for near a month,) it began to *snow* early in the afternoon, and there fell in the course of the ensuing night and day a considerable quantity, making 0·78 in the guage when melted. Now, from the 21st of the Twelfth Month (the shortest day) to the date of this snow, is sixty days; and from the date of the heavy snow with which this winter began (prematurely as we thought) at noon, on the 22d of Tenth Month, to the shortest day, is also sixty days. Should the winter terminate with this snow, which has been followed, after a thaw, by some night-frosts, it will have lasted, with the usual mild intervals included, an hundred and twenty days, beginning and ending at the same point of the Sun's declination. In this respect, then, our winter will have been coincident, for once, with that of a much higher latitude: and as we have been intruding of late years, with our ships of discovery, into the polar circle, the *North* may be said to have *returned the visit!*

To proceed.—Tab. 22 exhibits a good specimen of the Solstitial rains. The period begins with a SE wind, and a max. Temp. of 85 ˇ In five days, with 2·76 in. of rain, and thunder, it is lowered to 67°, the Temp. of the nights keeping up: the winds these five days as follow, S, NW, NE, E, W. The Evaporation, which is about 0·10 in. *per diem,* certainly proves nothing in favour of a *saturated* state of the lower air all this time. The NW, W, and SW prevailed on the few fair days in this period, which had 5·10 in. of rain. See also Tab. 47. 59.

In Tab. 25, we have several considerable amounts of rain brought by strong SW winds: while a little rain likewise attends the interruption of the course of NE immediately succeeding.

Tab. 31. Here we have a remarkable succession of the daily winds during rain, *and which proves introductory to fair weather.* First, Fourth month 14, *var,* with thunder and hail, then SW, W, NW, N, SE, NE, and, after a few days of changeable winds, the reverse order, NE, N, NW, W, SW, with a day *variable* at the commencement of the dry weather; which prevailed for the most part during the month after.

Tab. 42. In this period, at the beginning of Spring, we have a striking contrast between the effects of the Southerly and Northerly winds: a course of the for-

mer, with daily rain, giving place in the middle of the
period, to a course of the latter, with dry weather and
frequent hoar frosts.

Tab. 51. (Eleventh month 19 to Twelfth month 18,
1810.) We have here the enormous amount of 5.54
inches of rain in the space of 30 days, with appropriate winds, and an *electricity* which might have become the heats of summer. Whether from this cause,
or from the temperate warmth and moisture by night,
or both united, the *ignis fatuus,* a phenomenon scarce
known in this part of the island, appeared in the
marshes near our Laboratory, on several nights during a very wet week; and gave place only to the
overflowing of the river, which laid the ground under
water. I did not get to see this rare visitant, and am
consequently unable to speculate from actual observation upon its nature : but some circumstances which
were told me by an eye witness, respecting the brightness and swift gliding motion of the lights, induced
me to think them electrical ; and I am disposed to
class them in the present instance with the smaller
kind of *shooting stars,* though making their appearance in so very different a region of the atmosphere.
It is possible that the evolution of phosphuretted hy-
drogen gas may sometimes produce luminous phenomena in these situations ; but on this supposition they
ought to appear more frequently : lastly, to conclude
these conjectures, there are extant descriptions of
ignes fatui, which are scarcely to be explained on
any other hypothesis, than that some insect, with
which we are perhaps acquainted in its ordinary
appearance by day, becomes luminous when collected
into dense swarms, and flying thus by night.

There is annexed to the results of the preceding
period (Tab. 50) an extract from the Papers, in which

a meteor is described of such a nature that, supposing the relation true in all particulars, it would be difficult to bring it under any class of phenomena yet established. I gave this confused statement as I found it, to serve for the use of any observer in whose way such an appearance may chance to come hereafter.

Tab. 61. A contrast in the opposite season (if we include also a few days of 60) to the arrangement of winds and rain in Tab. 42. Here we have a course of dry Northerly and Easterly winds, followed by a series of Westerly gales with daily rain; the introduction of the latter being marked by *three* days of SE and S; and one *variable.*

The crossing of the currents, and the effects of their mutual contact in electrical precipitation, appear in some extracts from the Papers annexed to this period. See the Notes respecting the winds at *Plymouth* and *Harwich* : and compare with these the winds at *Plymouth* and *Hull,* under Tab. 67. Both cases being evidently connected with the Equinoctial season, and introductory to rains about equal in amount and duration.

Tab. 65, 66, and 74, exhibit a great number of instances of the connexion of the S and SE crossed by the NW, as also of the variable winds, with heavy rain, both in the early and latter part of the year. The usual electrical phenomena ensue, upon the copious decomposition of the vapour, in each season.

Tab. 69. In this period the gradual *southing* of the wind before rain is thus *twice* exhibited in its daily changes, 1. NE, E, SE, SW; then under different winds during a week, 0.60 in. rain. 2. N, NE, E, SW; during this, again, 0.61 in. rain, and immediately with *var.* for one day, 0.60 in. rain. Then, NW a day, *fair* : then, E, SE, SW; S, SE, SW;

S, SW; with 0.55 in. rain in 10 days; after which followed two weeks of dry weather with a high barometer.

Tab. 81 has a curious mixture of winds, with continued rains, in the Spring season; and, as usual, thunder: but it would be tedious to point out to the reader the many cases which he may find (if disposed to prosecute the enquiry) in these Tables.

Probable sources of the Vapour brought by different Winds.

The introduction of a surplus of vapour from the S or SE, and its decomposition by the prevalent NW, and in some cases the W and SW winds, will now be evident: as likewise the reason why the SE wind is so intimately connected with electrical indications, with hail and thunder.

Vapour brought to us by such a wind must have been generated in countries lying to the South and East of our island. It is therefore probably in the extensive vallies watered by the Meuse, the Moselle and the Rhine, if not from the more distant Elbe, with the Oder and Weser, that the water rises, in the midst of sunshine, which is soon afterwards to form *our* clouds, and pour down in *our* Thunder-showers. And this island, in all probability, does the same office for Ireland: nay, the Eastern for the Western counties. My attention was lately called to this subject by a striking fact, which occurred in preparing for the press the 131st Table of this volume. After nearly nine days wet weather, attended as usual with mixed winds, in our district, upon the wind changing from SE to NE, it became fair with us; and on the same

day (the 26th of Fifth Mo. 1817) a rain of three days
and nights commenced in the country East of the
Upper Rhine about *Stuttgard,* so heavy as to produce
a serious inundation. In the mean time we had no
rain, though the Barometer was still very low, and
the change of wind above mentioned had been attended with thunder. The rain ceased in those parts
upon the evening of the 28th, and on the next two
days it rained again with us. To suppose a connexion
of the phenomena at this distance on electrical principles may be too much: but I think one may be
made out through the medium of the winds, in this
manner. The evaporation of a tract of country lying
to the East of both stations, might in the first instance
be conveyed to the Thames, and then, by a change
in the direction of the prevailing wind, to the sources
of the Rhine: and decomposed into rain, with us by
the effect of a colder latitude ; and with them by that
of the elevation of the country, aided probably in both
cases by opposing currents.

Thus, drought and sunshine in one part of Europe
may be as necessary to the production of a wet season
in another, as it is on the great scale of the continents
of Africa and South America ; where the plains, during
one half of the year, are burnt up, to feed the springs
of the mountains; which in their turn contribute to
inundate the fertile vallies, and prepare them for a
luxuriant vegetation. And we may now be more able
to understand the unequal distribution of the wet
summer of 1816 ; when as I have already stated, under Tab. 122, the middle of Europe was suffering
from excessive rains, at the same time that the North,
or the parts East of the Baltic at least, about Dantzig
and Riga, were suffering from drought, and in all
probability furnishing the water.

In the Spring and Summer, both the direction of the winds, and the relative state of Temperature, seem to forbid our receiving much rain from the Atlantic. But in winter, when the surface of the ocean is giving out heat to the air, it may be supposed also to give out vapour, in greater quantities than the Temperature of the air is prepared to sustain. Hence the Atlantic, during the winter months, or rather in the interval between the Autumnal equinox and the Winter solstice, is probably the great source of our rains. The impetuous gales which, at this season move over its surface, and impinge on our Western shores, may *possibly* bring us much vapour from the superior atmosphere of the tropic where they originate. The powerful manifestations of Electricity which at times attend them, seem to favour this opinion. But should they have deposited much water on the passage, we may still find, in the relative winter temperatures of the air on our coasts and the ocean, a sufficient reason for the *turbid* state in which they are almost uniformly found on their arrival.

OF THE LUNAR PERIODS.

THE Variations of the winds and temperature have been shewn to have an intimate connexion with the Rains of our climate. There is another relation of the kind, much more generally attended to, and on which it might now be expected that I should say something; I mean, that of the Rain to the indications of the *Barometer*.

That the Barometer descends gradually before rain, and rises during or subsequent to it, and thus indicates the return of fair weather, is matter too trite for me to enlarge on here: and that the *mean* of the observations is higher in the dry and fair periods, and consequently in the years in which these predominate, than in the wet ones, is what every attentive reader must have found abundantly proved in the results which have been before him. The reason of this must even have become evident; and he will now scarcely need to be reminded, that the air *weighs more* when it is warmed and charged with transparent vapour to a great elevation by sunshine, than when, being chilled by the long nights of winter, it is shrouded in stormy clouds, and undergoing continual decomposition.

But furnished only with these general notices, he will find himself at a loss to explain many of the movements of the column: to know why it is generally high in severe frost, or with a North-east wind; and why sometimes very low without the expected accompaniment of much rain: in short, he will desire to account for those large sweeps which it makes occasionally, without any regular connexion with the changes of wind or weather; and for its apparent stagnation at other times, about a middle point of elevation, while the most obvious perturbation in the atmosphere is going on, and rain and thunder occur daily. Nor will the sudden depressions attending our southerly gales, and the rapid manner in which the former level is restored after them, escape his enquiry. A clew to the chief of these difficulties is furnished by the certainty, now sufficiently ascertained, that the atmosphere is subject, like the *liquid* ocean, to the attractive influence of the Moon, and from this cause, operating jointly with the Sun's power, it has *its tides and currents*. It was from the supposition of this fact, not indeed without some ground of observation, that I was induced to cast my Reports on the weather into the form of *Lunar periods*. I shall not undertake here to give the theory of the Lunar tides in the atmosphere. Indeed what I have hitherto learned respecting them, appears to constitute but an imperfect glimpse of this difficult subject; which will now possibly claim the attention of men duly qualified to investigate it; to whose service the materials to be found in these volumes, and which constitute a greater store than I can pretend to use, are cheerfully dedicated. They will be extracted however the most readily by those who will first condescend to go through the labour of proving, to themselves, the general correctness of the following details of my progress hitherto.

Influence of the Moon on the variable pressure of the Atmosphere, on the Temperature, Winds and Rain.

1. By the Moon's change of place in her orbit.

A series of observations which I made in the year 1798, on the variations of the Barometer in connexion with the Lunar phases, may be found, illustrated by a plate, in Tilloch's Magazine.*

In registering the movements of the Barometer at that time, I employed instead of figures a curve, traced from day to day on a graduated Chart, sold for that purpose by a Copper-plate printer in London. Finding by the specimen which accompanied these blank Registers, each of which served for a month, that the Moon's phases were to be inserted where they occurred, by an appropriate sign, at the top of the column for the day, I adopted the practice, not without some previous, though slight, information on the subject.

My observations had not proceeded far, before I perceived a coincidence of the greater *elevations* of the Barometer with the Moon's First and Third *quarters*, and of the greater *depressions* with the *New* and *Full* Moon. When the year had been completed on my charts, I gave an account of the subject to the *Askesian Society*; a select company which met every fortnight, at the house of my friend William Allen, in London, for the purpose of philosophical discussion. By this society they were favorably received and published : the substance of the paper was as follows:

In above thirty out of the fifty lunar weeks of that year, the curve reprsenting the movements of the Barometer changed its direction in such a way, as to

* Philo. Mag. vol. 7, p. 365, &c.

be either falling, or at its minimum for the space of
two weeks, under the phases of New and Full Moon;
and rising, or at its maximum for the like space, under the First and Third quarters.

The remainder of the year presented exceptions
sufficiently decided to forbid this coincidence being
taken for a general rule. The case was sometimes
indeed reversed, so that a low Barometer coincided
with the Quarters, and a high one with the Full and
New. And perceiving that the *rule* obtained chiefly
in moderate and settled weather, and the exceptions
when it was stormy, frosty or inclined to thunder, I
came to the conclusion, that the former mode of variation exhibited the regular Lunar Tides, and the latter, such a mixture of tide and currents as might be
expected to belong to a perturbed state of the atmosphere.

To ascertain the effect of each Lunar position, independently of the variations supposed to be produced
by currents, I took the following method : The height
of the column at the time of the occurrence of each
phase was taken, and the separate observations falling
under each class reduced to an average ; which average was then compared with that of the whole of the
observations thus taken, The mean of the whole was

29·9638 in.

Average at Full Moon   29·906
            which is less than the mean   ,0578 in.
        at Third Quarter 30·153
            which is more than the mean ,1892
        at New Moon   29·719
            which is less than the mean   ,2448
        at First Quarter   29·980
            which is more than the mean ,0162

The result of each position was thus found, on the whole, to agree with the *rule*; but was by no means in proportion to its occasional manifestations. For, in the course of the first seven weeks of the year, there appeared three great elevations in the curve, the summits of which were nearly coincident with the First or Third quarter; and one of them precisely so with the Third quarter, at the extraordinary height of 30·89 inches. The Barometer was the same from which my observations are still registered, and was in perfect order at the time. The *mean* of the season was however a very high one; so that this pyramid stood on an elevated base. On the other hand, the most remarkable depression of the year, to 28·60 in., occurred only twelve hours from the time of New Moon; and several other considerable depressions were nearly coincident with New or Full Moon.

I determined, upon this, to extend the enquiry, and I selected for the purpose the years from 1787 to 1796 inclusive, as they stood in the Register of the Royal Society. The results, obtained by the same method, were as follows.

Mean of the observations taken out

29·818 in.

Average at Full Moon   29·7812
      which is less than the mean   ,0368 in.
    at Third Quarter 29·8823
      which is more than the mean   ,0643
    at New Moon   29·7946
      which is less than the mean   ,0234
    at First Quarter 29·8910
      which is more than the mean   ,0730

It was not to be doubted that these numbers presented a more correct scale of the effects, than could be expected from any single year's observations. Assuming therefore the elevation of 29·818 as the standard of comparison, I concluded that the Barometer is depressed, on an average, about a tenth of an inch by the change of the Moon's position from either quarter to the Full or New, and elevated in the same proportion by her return to the quarters. It would have been more correct to have taken for the standard the mean of the whole ten years observations in that Register, or 29·89 inches; and to have stated the effects thus : the Barometer, on an average of ten years at London, suffers a depression of about a tenth of an inch, by the influence of New and Full Moon respectively : but at the First and Third quarters the Moon's influence is, in respect of position in her orbit, neutral, producing neither elevation nor depression in the Barometer.

Having satisfied myself as to the fact of an influence of the Moon upon the variable pressure of the atmosphere, I proceeded to draw a parallel between this case and the tides in the Ocean, thus : " I suppose therefore, that the joint attraction of the Sun and Moon at New Moon, and the attraction of the Sun predominating over the Moon's weaker attraction at the Full, tend to depress the Barometer, by taking off from the weight of the atmosphere (its counterpoise) as they produce a high tide in the waters by taking off from their gravity : and that the attraction of the Moon, neutralised at the quarters by that of the Sun, tends to make a high Barometer, together with a low tide in the waters, by permitting each fluid to press with additional gravity upon the Earth."

An objection was then anticipated which might arise from the circumstance of the *diurnal* tides being the most considerable in the ocean, and the weekly

elevations and depressions contributing, as they proceed, but a moderate proportion to each day's tide :
whereas, in the *atmosphere* of these latitudes, where
the weekly elevations and depressions go to so large
an extent, the diurnal tide is scarce perceptible. This
objection was attempted to be met by some reasons,
founded on the very different physical constitutions of
the air and ocean ; the latter being pretty uniform in
density and composition, while the former is variable
in these respects, and subject moreover, by the rapid
changes of temperature which it undergoes, to currents
which move in different directions with much greater
freedom than in the ocean. It was also stated that at
Calcutta, where the weekly variation, as in low latitudes generally, is very small, a daily tide had been
distinctly traced by the alternate elevations and depressions of the Barometer. I have also since the date
of this essay, met with strong indications of a daily
tide in Registers of a much higher latitude.

This paper had in substance the following conclusion : " It will be soon enough, however, to enter
upon the Theory of the Atmospherical tides, when
the facts shall have been examined, and the influence
of the Sun and Moon on the atmosphere established
by more extensive observations. For this purpose
the subject is now brought forward, and the co-opera-
tion of observers in this or other countries is requested.
The coincidence, as far as hitherto observed is an
important fact, and should it be found to obtain generally, will lead to important consequences, and in the
first place to a new and more satisfactory theory of
the Barometer. The true reason, likewise, of the
weather so frequently agreeing in its changes with
those of the Moon (a coincidence which has long
served to direct the predictions of the Almanac-

makers) will be apparent; and the meteorologist will
avail himself of this, to form probable conjectures on
the changes likely to arise for a certain time, not
exceeding that which limits the operation of the known
cause or causes."

Such was the state of my information on this subject
twenty years ago The study of the modifications of
clouds, and the various phenomena connected with
them, afterwards occupying my attention, this particular enquiry was suspended : and when, in 1806, I
began a regular Meteorological Journal, it was with
more general views. But being still desirous of putting to the test the opinion of a Lunar tide in the
Atmosphere, I was induced, as already mentioned, to
digest the observations in the form which seemed to
afford the greatest facility for this purpose. It remains
to show how far the purpose was fulfilled.

———————

Plate 4 exhibits, in a system of curves, the variation of the *daily mean height* of the Barometer through
the *Solar year* 1806—7. These curves are constructed
from the Tables, 2 to 14 inclusive, in vol. 1, but with
a different arrangement of the periods. For the sake
of shewing more evidently the influence of New and
Full Moon, the periods are here made to begin with
the day of the third quarter; which happens in this
instance to be the first day of 1807. After a dotted
curve therefore, giving the variation from Dec. 22,
1806 to the end of that year, the several curves *a—b*,
b—c, &c. carry on a series of entire Lunar periods to
the 21st of Dec. 1807, at *n*, where the Solar year closes.

To construct these curves, the *mean of the period* is
first ascertained by calculation, and represented by a

horizontal line. The relation of the *mean of each day* to this standard line is next ascertained and marked; and a curve carried through the points thus found, represents the variation, at its proper extent, above and below the standard. In doing this, the mean of the day on which a Lunar phase happens, is made always to fall in the intersection of the curve with a perpendicular line appropriated to that phase. Some inequalities of time in the intermediate parts of the curve, occasioned by this arrangement, are remedied, where needful, by using an unequal scale of time in those parts.

By this method the curves were all obtained of an equal length, and presenting equably the relation of the Lunar points to the Barometrical mean for the period. Their tendency to rise and fall at particular intervals, and their consent or opposition in such movements, was thus also represented *independently of the absolute place of the mean of the period, or of that of each day, in the Barometrical scale.* Each of the four horizontal lines, on which the curves are made to play, has therefore an elevation peculiar to itself, and relative only to the curve in connexion with which it is viewed : its absolute place in the scale of the Barometer may be gathered from the small curve at the bottom of the diagram ; where these monthly means are laid down upon the mean of this Solar year, which is 29.815 inches.

All this contrivance was needful in order to exhibit the distinct effect of each Lunar position, unmixed with that kind of variation, from month to month, in the mean of the Barometer, which depends on the season of the year ; and of which an account has been given page 150—152, founded, it will be recollected, on a mean of several years.

To proceed now to the application—it is difficult not to be struck at first sight with the evident marks of *system,* which these curves exhibit, from the beginning to the end of the series. Were it possible to obtain, at successive equal intervals of time, the profile of the waves which roll after each other on the surface of the ocean, and were we to reduce these to a scale in like manner, it is not to be doubted that the group would present elevations and depressions indifferently, in all parts of the scale of time; and the intersections of the curves would soon produce confusion in the picture. But it is not so here—the wave occurs too often in the same place; and the intermediate depressions are too regular, for us to admit, that what is called *chance* has any considerable share in producing them.

The most prominent feature of the piece may be said to be, the nearly constant *elevation of the curve* at the approach of *Full Moon*—a very contrary result, certainly, to that found in the year 1798, and sufficient, at first view, to invalidate the partial conclusion I then came to, that the true atmospherical tide consisted, in part, of large depressions at this quarter. These elevations, however, will be found to have their *apex,* for the most part about two days before the Full, and to be going off at the time of the phasis. That they are properly connected with its approach, may be fairly inferred, from the manner in which the curve No. 2 rises at this time from a great depression, as if prevented from taking an upward tendency by some unusual cause, and become more elastic in consequence of being thus strongly bent downward.

If we now turn to the *New Moon,* on the left of the plate, we perceive its approach marked, by depressions chiefly in the fore part of the year, and by ele-

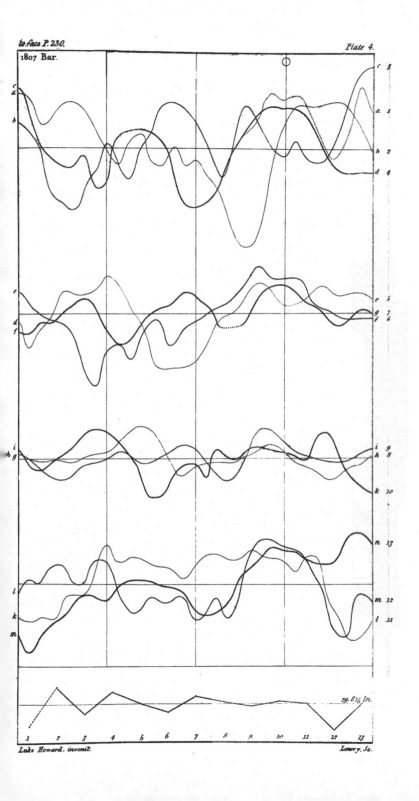

to face P. 230.   Plate 4.

1807 Bar.

29.815 In.

Luke Howard, invenit.   Lowry, Sc.

vations in the latter part. Yet the actual time of this position, or rather a day or two after it, exhibits a strong tendency in the diurnal variation to return to the mean of the period: and the same observation applies to both the other *quarters*; which have also some peculiar *opposite* variations connected with them. The latter are conspicuous in the elevations which belong to the *third quarter* of the first three periods, and in the depressions which attach to it in the last three: each however with an exception attached: see Nos. 4 and 13.

Enough has perhaps been pointed out, to satisfy the reader that in this year, there was a decided connexion between the Lunar positions, and the mean daily movements of the Barometer; which deviated in the same direction about the same point of a Lunar revolution, whether the mean of the season occupied the higher, lower, or middle part of the scale.

A certain relation was long since found to obtain between the movements of the Barometer and the variations of Temperature in the atmosphere: and very early in the course of these enquiries I perceived, on tracing the curve of the diurnal mean temperature on the same scale, and referable to the same mean line, with that of the Barometer, that the connexion was almost constant between them. It is manifested in two different ways, which may be termed conjunction and opposition; since in the one, the curve of the mean temperature accompanies (or precedes or follows by a short interval) that of the Barometer, and in the other the two vary in opposite directions, often with a very near coincidence in time. See Fig. 12, p. 235.

Two degrees of Fahrenheit are equivalent in these variations to a tenth of an inch in the Barometer. Such are the proportions observed in this figure, the parts of which are copied from some of the many periods I have traced in this way. When the two curves run in opposition through a period, they cross at intervals, and form a succession of rhombs, differing in magnitude according to the extent of variation in either or both of the curves: when the two run in conjunction, the resemblance in the number and extent of the changes is often so close that the one might easily be mistaken for the other. There are also many periods in which both the kinds of relation appear; and some in which neither is very obvious.

In Plate 5, the variations of the *daily mean temperature* through the Solar year 1806–7 are traced in curves, bearing the proportion already mentioned to those of the Barometer, and constructed in other respects on precisely the same plan as in Plate 4. The corresponding numbers on the curves in each Plate will serve to connect those of the Temperature with the Barometrical ones for the same periods.

The place of the *mean line of each period* in the Thermometrical scale, is indicated in the curve at the bottom of the plate.

These curves present features in some respects less striking than those of the Barometrical variation; but which, when attentively examined, indicate equally the existence of a system of variations, governed by the Moon's attraction, as a secondary cause, subject on the whole of the year, to the more powerful influence of the Sun as he varies in declination.

The greater variations of Temperature, it may be first remarked, appear for the most part during this year in the intervals of the lunar phases: and there is

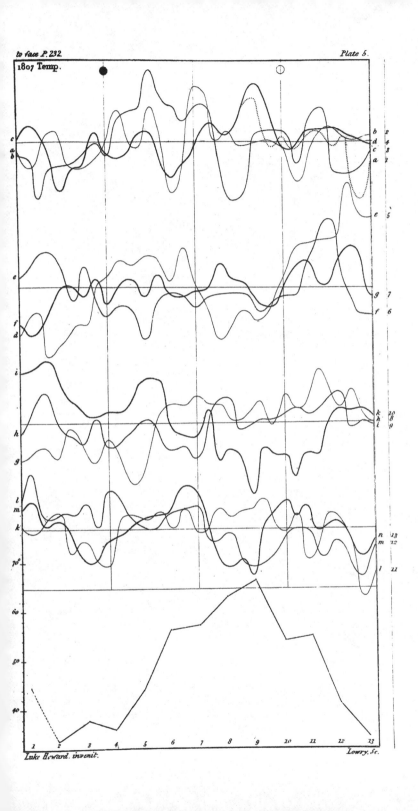

Plate 5.

1807 Temp.

Luke Howard, invenit.

Lowry, Sc.

a tendency in the curves to return about the time of
the phases to the mean line, commonly in order to
cross it, and assume an opposite deviation; from
which they often return within the week, as before.
The change of the mean of the period, again, from a
lower to a higher place in the scale, or *vice versa*,
according to the season, is effected not so much by
the gradual elevation or depression of the Temperature through the period, as by sudden bold sweeps of
the curve in particular parts of it. Numbers 7 and 8
for the summer, and 10 and 12 for the autumn, approach to the former, or gradual mode of variation;
while 5 and 11 may be cited as instances of a more
rapid change of level. In each of the latter three
periods, the curve assumes a decided tendency upward
or downward, two or three days before the Full Moon,
which it preserves through the following week, the
warm or cold weather coming in at once by means of
this movement. No. 3 and 5 present almost equally
bold upward sweeps, having their *nodes* (if I may be
allowed so to use the term) about New Moon ; but
these elevations do not hold their level afterwards.

If we contemplate the cold periods, No. 2, 3, and 4
in connexion, their general character, notwithstanding
a large depression in each about the middle, will
appear to be that of a rising Temperature by the influence of the *New,* and a falling one by that of *Full*
Moon.

In the periods from No. 5 to 8, of increasing Temperature, the near agreement in the time of beginning
their most considerable elevations above the mean
will scarcely be thought accidental. Period   takes
its departure from the mean of 4 and closes very little
above that of 6 : this period has hence in effect five
points of intersection with the line ; which limit four

distinct and contrary oscillations of Temperature, each performed in the space between two Lunar phases.

Lastly, in the three periods of descending Temperature, No. 11, 12 and 13, there are six or seven depressions nearly coincident with each other in time. I can scarcely omit to notice here the beautiful manner in which the curve of the mean Temperature (like that of the Barometer) sometimes proceeds in gradually increasing and decreasing oscillations, about a general level or line of direction, which it has assumed for a few days. Period 8 has two examples of this, one below the mean line, the other above it: by the latter the Temperature was carried, on the 22d of the Seventh Month, to the higher extreme of the year: and No. 11 presents a third, in the course of which three weeks of fine weather (which had been attended with an appropriate variation of both instruments) broke up, and gave place to the Autumnal rains.

The Barometrical variation will be found, on comparing together the two systems of curves, to be mostly in opposition to, but at times in conjunction with, the Temperature. In the early cold periods, and in the fine weather of summer, *opposition* will be found predominant; but in the decline of the year, when the atmosphere is losing both heat and water, the two curves often vary in the same direction.

Fig. 12 contains specimens of Barometrical and Thermometrical curves in each state of relation. In the first pair, the season being frosty (the time, the first ten days of 1807) the Barometer ranges high, yet descends a little, to meet an elevation of the Temperature above the mean of the period in the first week. After this, with a South wind, the two curves suddenly change places, marking an intermission of short continuance in the frost.

Fig. 12.

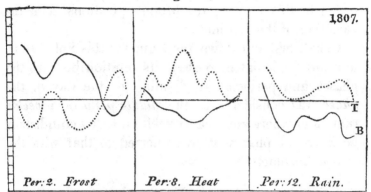

Per: 2. *Frost*   Per: 8. *Heat*   Per: 12. *Rain.*   1807.

In dry hot weather we have the reverse of this arrangement: the Temperature forming oscillations above the mean, and the Barometer an opposite curve below it. Such is the character of the variation for the space of eight days following the 18th of the Seventh Month, chiefly included in period 8, and represented by the second pair.

The third pair is a specimen of the agreement in direction of the two curves, when the season is tending to rain. Here we have the Temperature above the mean, but descending; and the Barometer below it, descending also: a slight opposite movement being felt, at the same time by both instruments. This specimen is a part of Period 12, beginning the 1st of the Eleventh Month, and is by no means the most interesting example which my set of curves, as far as already made out, would furnish. This week furnished about an inch of water, to the rain-guage at forty-three feet elevation.

———

We have next to enquire into the connexion of these variations with the changes of wind, and distribution of rain in each period; which will be

found strikingly unequal, and quite as much influenced in this year by the Moon's positions, as is the
variation of the Barometer.

I shall first put down the Rain for this Solar year,
in a form calculated to shew its relation both to the
phases and periods. In dividing it, the day of the
phase was considered as the *middle point* of a *week's*
rain; and where any quantity fell on a day equidistant
between two phases, it was referred to that with the
lowest Barometer.

Period.	Last Qr.	New M.	First Qr.	Full M.	Last Qr.
1.	...	....	....	0·04	0·05 in,
2.	...	...	0·27	0·21	....
3.	0·39	0·31		0·24	0·01
4.	0·01	0·25	0·01	0·02	...
5.	0·37	....	0·17	....	0·04
6.	....	1·57	0·79	....	....
7.	0·82	1·22	....	....	....
8.	....	...	0·05	...	0·01
9.	0·25	0·08	1·09	0·01	0·03
10.	0·28	0.05	0·65	...	0·41
11.	0·41	0·13	00·2	....	0·52
12.	0·46	0·21	1·42	0·11	0·49
13.	0·64	....	0·16	...	....
Totals	3·63	3·82	4·63	0·63	1 56 ⎫ 3·63 ⎬
					5·19

Add one-third for the level, the guage being at an elevation of
43 feet ...   .1·27   1·54   0·21   1·73

Rain at the ground   5·09   6·17   0·84   6·92

Total for the Solar year....19·02 inches

The great and almost positive *dryness* of the *Full
Moon week* during this year, is thus rendered equally
conspicuous with the elevations of the B rometrical
curve by which it was accompanied.

The immediate cause of both will presently be shown to be, *the prevalence of Northerly winds during this part of the Lunar revolution.*

A space of eight days being taken out of each period, for the New and Full Moon respectively, with the phase as nearly as could be in the middle of the time, the daily observations on the wind were found to number as follows:

Winds.	New M.	Full M.
N and NE	13	20
N-West	7	21
West	19	15
S-West	33	17
S and SE	6	4
East	5	11
Totals	83	88

The Northwest, which has been already shewn to be more peculiarly our fair weather wind, appears here thrice under the aspect of the Full, for once under that of New Moon: and the North and Northeast are more frequent in the former, in the proportion of 3 to 2.

On the other hand the New Moon, which exhibits so many depressions of the curve, has about double the number of South west, and a proportion of 3 to 2 of Southeast winds.

The West wind predominates here in the division allotted to the New; and the East, to more than double, in that of the Full Moon. I do not consider this disparity as so much connected with wet and dry as the former between the Northerly and Southerly winds. But so far as it is concerned, the *East* wind appears to have been productive rather of *rain*, as will appear by the following statement, in which a

week's observations are taken out, for each quarter,
with the phase in the midst.

Winds.	First Qr.	Third Qr.
N and NE	8....	21
N-West	.. 9...	6
West.......	...20..	10
S-West.... ...	.. .26.	13
S and SE ..	...... 2.	8
E	.... 0..........	9
Totals	65	67

Here the Third quarter, which is the wettest phase,
has thrice the amount of Easterly winds that appears
in the First: and only half as much Southwest. But
in a very dry year it is not so easy to decide from
what quarter we receive the rains, as when the cases
of heavy rain are multiplied.

On the whole of this year, a connection between
considerable depressions of the Barometer and the
more copious rains is sufficiently apparent; although
there are large depressions attended with but little
rain. For the former, see periods 3, 6, 10, 11, 12, 13:
and for the latter, 2, 5. In periods 3, 4, 7 and 9, there
are examples of rain connected with a mean height of
the Barometer, and a mean Temperature for the season.

The influence of the Moon on the Temperature and
density of our local atmosphere, appears therefore, with
respect to these more obvious and frequent changes,
to be exercised chiefly through the medium of the
winds. It is a secondary effect of her varied attraction; which continually tends to change the bearings
of the different currents in motion in the great body
of the atmosphere; and we are thus successively in-
volved in all their modifications. Not but that there

are seasons when the predominant Solar influence is exerted to a degree, which renders these Lunar changes of small consequence : and when in spite of the various aspects of our attendant planet, we are drenched with rain or parched with drought, for months together.*

* " By what law of Nature is the atmosphere governed ? We have not had any rain, generally speaking, since last harvest. (date, Feb. 19, 1777.) Springs have not yet begun to rise, deep wells in general want water, and many ponds are not yet filled : even the surface of the earth is not satisfied." Marshall : Minutes, &c. on Agriculture in the Southern counties.

The same.—" June 23. The spring seed time was moist, but not remarkably wet : the clouds reserved their bounty for May and June. The middle of May was very wet, and so is the middle of June. The last ten days have been, except one, uniformly rainy. Last night, it poured for eight or nine hours : perhaps never more rain fell in so short a time ; the ground was never so wet since the deluge !"

The same.—" July 15. From 23d June to the 8th inst. there was scarcely a fair day. The rain set in June 13 : it therefore lasted 26 days, with scarcely one fair day intervening. The attendant circumstances were these. The Barometer *hovered about changeable,* and seemed to watch the motions of the wind, which was generally SW. Whenever it veered round to the Northward, the air got heavy ; but as soon as it returned to its old station, the Barometer as regularly got back to changeable.

" *The impotence of the Moon was fully proved: she became full, shifted her quarters, and even changed, without the least effect.* The wind alone seemed to rule : for as soon as it was fixed in the North, the rain ceased, and before it had been eight and forty hours there, the weather changed from very cold, for the time of year, to very hot. The change of the wind was preceded by a very heavy squall in the night."

On these facts I would remark as follows. The law by which the atmosphere was on this occasion governed, appears clearly to have been the ordinary law of compensation. A long dry time preceded a long wet one : and the distribution of wet and dry, instead of being comprised within a month, (as is often the case with us,) occupied three whole seasons ; the dry extending from the Autumnal Equinox

----

The variation of 1807, like that of 1798, appears to be in great measure peculiar to the year in which it is found, and it gives place in the succeeding years to a different set of combinations.

The elevations of the *Barometer* about the Full Moon, for instance, which appear in 1807, are found in much less proportion in the next year; and in 1809 they mostly yield to depressions in the same place, the New Moon acquiring in the mean time larger and more numerous elevations.

With regard to *Temperature*, again, the different positions afford different results as the years proceed. In 1807, the average of the mean Temperature taken upon each day through the twelve periods, exhibits a very regular appearance. The Temperature thus obtained being laid down in a curve upon the mean of the whole, it is found to descend below the mean line, in the intervals between Last Quarter and New Moon, and First Quarter and Full Moon respectively, the depressions being carried a little beyond the latter phases: it then rises more abruptly than it fell, and the elevations thus formed in the alternate intervals go off before the arrival of the Quarters. But in the two following years, the parts occupied by these elevations were found by the same method to be passing

to the Vernal; the wet from the latter to near the point of highest temperature, a month after the Summer solstice: when the Southerly current suddenly shifted its range, and we were again placed in the dry air returning from the Northward; which, together with a clear atmosphere above, brought on a free radiation and warm weather. " The impotence of the Moon" during the rains, appears to have been a consequence of the absolute controul of the Sun over these currents through the season.

off into depressions, and those occupied by the depressions first rising to the mean line, and then becoming elevations.

The mean Temperature of these respective intervals for 1807, taken at equal distances and with a clear day allowed after each phase, were found as follows:

Mean Temp. from Last Quarter to New Moon 47.04°

New Moon to First Quarter 49.66°

First Quarter to Full Moon 47.67°

Full Moon to Last Quarter 49.78°

The *proportions*, only, of the rise and fall would have been somewhat different, had the Temperature been taken strictly from phase to phase. The Temperature of our atmosphere during this year was therefore alternately elevated and depressed to the amount of at least *two degrees* in each Lunar week, by some cause connected with the Moon's positions : which yet did not operate precisely in the same way in the following year. Indeed the curve of the Lunar mean Temperature for 1809, obtained in the manner before mentioned, is in its general appearance a contrast to that of 1807

The Full Moon week also loses in 1808, its *dry* character ; which is not immediately taken up by another phase : it exhibits in this year about 4 inches of rain ; and rather more in 1809. The *wet* phase in 1808 is the First Quarter ; and it is so again, though with a smaller excess over the other quarters, in 1809 : the Last Quarter becomes drier in proportion.

The relative changes in the direction of the prevailing winds, in each part of the Lunar period, for these two years, have been as yet but imperfectly examined.

A great depression of the Barometer appears in 1807, in the period No. 2, which goes off with a remarkable upward sweep of the curve, about the time of Full Moon. There are nearly parallel depressions, equally conspicuous, in the two following years. In 1808, the sudden rise after the crisis occurs 20 days earlier in the year, and with a like relation to the First Quarter: in 1809, it is about 20 days later, and attached in like manner to the time of New Moon. It is remarkable that in each case the full pressure was restored chiefly by means of *South-west* winds, and without any excess of rain, or storm of wind. Such periodical large movements, with the backward order of the phases in this instance, deserve notice; as being probably connected with extensive changes in our Northern atmosphere; perhaps with the shifting, through several degrees of longitude and latitude, of the range of the larger currents, which depend on the Sun's progress in North or South declination.

———————

Being curious to know whether the difference of the Lunar positions, which occasioned so unequal a distribution of the rain, had a similar effect on the *Evaporation*, I took out weekly portions from my Tables, with the phase in the midst of the time, as before for the rain; and found that even in 1807 and the latter part of 1806, the amount of Evaporation for 12 periods under the Full Moon was 9.84 in. the same under the Last Quarter being 9.55 inches. And having formed an average, for the three years of which I have more particularly treated, the amounts raised in equal times, under each phase, were found so nearly alike, as to render the conclusion inevitable, *that the Lunar*

*positions, however they may affect the distribution of
the rain, produce no sensible difference in that of the
Evaporation.*

This process is nearly a continuous one through the
year: it is an effect of the temperature of the water,
modified by the greater or less velocity of the wind
agitating its surface, and diffusing the vapour produced. But rain is an occasional process; and appears
to require a more complex arrangement of causes,
at least for its prevalence in a given district. We
have here to take into account the Temperature and
Electricity, absolute and relative, of both the earth
and atmosphere; the relative temperature, moisture,
and perhaps electricity of different simultaneous currents; the direction of these with regard to neighbouring seas and continents, and to the slope or exposure of the district itself; and lastly, as it seems,
the Moon's influence.

———— ———

I have now to give some account of this influence
as exercised on our atmosphere, *2ndly, according to
the Moon's place in North or South declination.*
The inquiry into this part of the subject was first
proposed to me by Silvanus Bevan, junior, of London, lately deceased. What I shall offer upon it
is principally derived from his minute and accurate
examination of the data furnished by my Register.
Other parts of the work had been before improved
by his assistance; and the diagrams were nearly all
finally prepared by his hand for the engraver. Had
his life been prolonged, I should not have been satisfied to conceal the obligations thus contracted to my
affectionate friend and zealous coadjutor; on whom a

large natural capacity matured by study and practice, had conferred great correctness of taste and judgment; and (what is yet a more pleasing reflection) his mental qualities were enhanced by the faith of a Christian, by an unblemished conduct and polished manners. His bodily constitution was however so feeble, that the utmost care over it sufficed not quite to middle age; and at the approach of the late winter, a pulmonary complaint, before habitual, became exacerbated to a degree which speedily brought on his dissolution.

The object of this inquiry, which my deceased friend had left imperfect, may be thus stated. Since it is evident that the Moon exerts an influence, through the medium of the winds if not also directly, on the atmosphere of these latitudes, the effects ought to be felt in a greater degree when that planet, by acquiring her highest North declination, becomes at her meridian altitude almost vertical to us, than when, being South of the Equator, she is vertical to a distant latitude in the other hemisphere. To ascertain this, it was necessary to submit some part of the observations, in my first volume, to the like test as in the case of the Lunar phases; by comparing, in detail, particular results with a general average. The years 1807 and 1816, the one the driest, the other the wettest of a series of 18 years, were selected as first entitled to notice; and the results have proved of greater value than either of us had anticipated. It is evident, from these two years alone, that not only the variable pressure of our atmosphere, but its mean temperature likewise, and the periods of the deposition of rain, are modified in those latitudes by the Moon's declination. Thus, another important feature is added to this already complex subject: and the

same anomaly, arising from the combination of the
different causes producing the phenomena, is found
here also—that particular results appear in opposition
to a general rule ; which rule is yet in the end satisfactorily established by general averages.

*Barometrical Averages, in Half-periods of Lunar
declination :* from 29–30 Dec. 1806 to 20 Dec. 1807,
or $355\frac{1}{2}$ days : Mean of the whole 29·816 inches.

Per.	Days	Moon	South	Days	Moon	North	Days	Mean of both
1	14	30·178	+221	13½	29·732	—225	27½	29·957 in.
2	14	29·676	—089	13	29·860	+095	27	29·765
3	14	29·928	+018	13½	29·892	—018	27½	29·910
4	14	29·907	+105	13½	29·694	—108	27½	29·802
5	13½	29·970	+246	14	29·486	—238	27½	29·724
6	13½	29·927	+044	13½	29·837	—046	27	29·883
7	13½	29·972	+026	14	29·921	—025	27½	29·946
8	13½	29·789	+022	13½	29·744	—023	27	29·767
9	14	29·847	+009	13½	29·866	—010	27½	29·856
10	13½	29·814	+056	14	29·705	—053	27½	29·758
11	13½	30·005	+136	14	29·738	—131	27½	29·869
12	13	29·486	—036	14	29·556	+034	27	29·522
13	13½	29·648	—206	14	30·042	+188	27½	29·854
Mean		29·857		Mean	29·775		Mean	29·816

Note. The spaces taken are those during which the Moon was
successively in N, and S declination : the fourth and seventh columns
shew the quantities by which the average height of the Barometer for
those spaces fell short of, or exceeded the average of the period, as
given in the last column.

Averages of the Barometer and Thermometer in Quarter-periods of Lunar declination, from the 3d of 1st Mo (Jan.) to the 23d of 12th Mo. (Dec.) 1807, or 355 days.

Mean Temperature 48·58°.

		Full South Declination			Mean Declin. Moongoing N			
Per.	Days	Barometer.		Therm.	Days	Barometer.		Therm.
1	7	30·203	+ ·389	33·36	7	29·938	+ 124	34·69°
2	7	29.434	— ·380	33·25	7	29·687	— ·127	43·32
3	7	30·161	+ 347	35·03	7	29·810	— ·004	32·14
4	7	29·691	— ·123	37·93	7	29·997	+ ·183	37·54
5	7	30·016	+ ·202	54·50	7	29·629	— ·185	60·36
6	7	29·943	+ ·129	60·36	6	29·721	— ·093	51·88
7	7	30·024	+ ·210	58·61	6	29·866	+ ·052	61·92
8	7	29·844	+ ·030	66·32	7	29·708	— ·106	69·32
9	7	29·792	— ·022	65·75	6	29·887	+ ·073	79·29
10	7	29;759	— ·055	54·07	6	29·903	+ ·089	15·83
11	7	29·786	+ ·172	58·46	7	30·071	+ ·257	58·39
12	6	29·510	— ·304	44·21	7	29·391	— ·423	39·69
13	7	29·808	— ·006	32·39	6	29·661	— 153	33·54
	90	29·852	Means	48·57	86	29·789	Means	49·57

N.B. The spaces taken are as nearly as possible those which have the Moon's greatest N or S declination, or her position on the Equator in their middle. The differences of the Barometer refer in this Table to the general average only, or 29·814 in.

Mean of the Barometer 29·814 in.

Full North Declination.			Mean Declin. Moon going S.		
Days	Barometer.	Therm.	Days	Barometer.	Therm.
7	29·375 — ·439	36·39	6	30·392 + ·422	33·50°
7	29·906 + ·092	38·93	6	29·699 — ·115	39·83
7	29·789 — ·025	34·96	7	30·102 + ·292	38·04
7	29·501 — ·313	48·82	6	29·704 — ·107	38·39
7	29·388 — 426	52·25	7	30·051 + ·137	54·36
7	29·822 + ·008	57·93	7	29·954 + ·140	58·64
7	29·844 + ·030	58·25	7	29·945 + ·131	61·14
7	29.728 — ·086	65·79	7	29·844 + ·030	62·86
7	29·806 — ·008	57 43	7	29·929 + ·115	62·07
7	29·714 — ·100	55·04	7	29·740 — ·074	53·11
7	29·667 — 147	53·32	7	29·598 — ·216	45·57
7	29·743 — ·071	37·54	7	29·316 — ·498	35·39
7	30·125 + 311	36·04	7	30·202 + ·388	32·18
91	29·724 Means	48·66	88	29·881 Means	47·53

In the Table, page 245, the Barometrical observations for $355\frac{1}{2}$ days of 1807 are reduced to averages,
on half-periods of 13, $13\frac{1}{2}$ or 14 days; during which
the Moon was in North or South declination. These
are contrasted, in each case, with the mean of the
whole period. In *ten* out of thirteen cases, the Barometer averaged above the mean, while the Moon was
in South declination; and below it, while she was in
North declination: *three* exceptions appear, which
belong to the *winter*.

The total results are these,

On $177\frac{1}{2}$ days with the Moon South,   29·857
On 178 days with the Moon North   29·775
Mean of the $355\frac{1}{2}$ days   . . .   29·816 in.
Elevation for her position South of
    the Equator . . . .   ·041
Depression for her position North .   ·041

A similar calculation of averages having been made
for 356 days, from the 24th of 12th Mo. 1815, to the
13th of the same, 1816, but without descending to
half-days in dividing the periods, the results are as
follows:

On 180 days, Moon South,   . .   29·765
On 176 days, Moon North,   . . .   29·704
Mean of the 356 days, ...   29·735 in.
Elevation for her position South   .   ·030
Depression for her position North   .   ·031

The Barometer having stood lower and ranged less
in this year than in 1807, the variation for declination is
less in amount accordingly. The cases which appear
against the general rule, or in which the Barometer
averages *higher* under a *North* declination, form in
this year a majority, occurring in seven out of thirteen

periods; and of these seven, *five* clearly belong to the *summer* half-year.

In the Table, page 246–7, the mean Temperature is taken along with the mean height of the Barometer for 1807, and each period is divided into *quarters*. The intention of this was, to ascertain separately the respective effects of a full South, of a full North, and of each kind of mean declination. In making up the results, the *Rain* for each of these quarters, ascertained by a separate calculation and corrected for the elevation of the guage, is likewise inserted. The results are,

1. For the quarter period in which the Moon was in Full South declination : Barometer   29·852 being above the general mean   . .   ·038 in.
Thermometer . . .   48·57°, being below the general mean ·01°.
Rain 3·56 inches.

2. For the quarter period in which the Moon was coming North across the Equator, Bar. 29·789 below the general mean . .   . . ·025 in.
Thermometer   . .   49·57°, above the general mean .   1·00°.
Rain 4·96 inches.

3. For the quarter-period in which the Moon was in Full North declination, Barometer   29·724 below the general mean .   . . .   ·090 in.
Thermometer . . . . 48·66°, above the general mean .   ·08°.
Rain 6·67 inches.

4. For the quarter-period in which the Moon was going South across the Equator, Barom. 29·881 above the general mean . . . . ·067 in.
Thermometer . . . . 47·53°
below the general mean · 1·05·
Rain 3·72 inches.

Having constructed a similar Table for a space of 355 days, beginning the 28th of the 12th Mo. 1815, and ending the 17th of the same, 1816, I found the results as follows; the general mean of the Barometer being 29·723 inches; of the Thermometer 47·09°; the Rain taken at the level of the ground.

1st Quarter-period, Barometer 29·797
above the general mean . . . ·074 in.
Thermometer . . . 46·14·,
below the general mean   0·95°.
Rain 6·65 inches.

2d Quarter-period, Barometer 29·793
above the general mean . . - ·070 in.
Thermometer . . 48·73°,
above the general mean   1·64°.
Rain 8·21 inches.

3d Quarter-period, Barometer 29·559
below the general mean . . . ·164 in.
Thermometer . . 47·00°
below the general mean   0·09·
Rain 9·99 inches.

4th Quarter-period, Barometer 29·678
below the general mean . . . . 0·55 in.
Thermometer . . . 46·51·
below the general mean   0·58·
Rain 5·49 inches.

Summary of the effects in these two years.

1. With the Moon Full South.

		1807			1816	
Barometer			above mean falling			above mean, rising.
Temperature			about mean			at lowest average.
Rain			the minimum quantity			near the minimum quantity.

2. With the Moon coming North.

		1807			1816	
Barometer			below mean			at highest average.
Temperature			at highest average			at highest average.
Rain			much increased			much increased.

3. With the Moon full North.

		1807			1816	
Barometer			at lowest average			at lowest average.
Temperature			about mean, falling			about mean.
Rain			the maximum quantity			the maximum quantity.

4. With the Moon going South.

		1807			1816	
Barometer			at highest average			below mean rising.
Temperature			at lowest average			below mean.
Rain			nearly at the minimum			the minimum quantity.

The most considerable and striking effect of the Moon's positions in declination here exhibited, is certainly that of the unequal distribution of the *Rain:* which I shall therefore first notice.

It appears that, while the Moon is far South of the Equator, there falls but a moderate quantity of rain with us; that, while she is crossing the Equator towards these latitudes, our rain increases; that the greatest depth of rain falls, with us, in the week in which she is in Full North declination, or most nearly vertical to these latitudes; and that during her return over the Equator to the South, the rain is reduced to its minimum quantity *And this distribution obtains in very nearly the same proportions both in an extremely dry, and in an extremely we season.*

The next point to be attended to is the Temperature, in which the two years exhibit (in this respect,) some striking coincidences.

In both years, the Temperature is at its highest average (for the period,) while the Moon is coming North over the Equator. During her continuance in North declination, the temperature in both passes the mean of the period, descending. In the *dry* year, it attains its lowest average while she is proceeding South again : but in the *wet* year, this takes place in the following week, or while she is in full South declination.

I have already exhibited for the year 1807, an unequal distribution of rain, as well as a periodical variation of Temperature, connected with the Moon's *phases.* It will be proper for the reader's satisfaction to recur to these, and to shew that both in 1807 and 1816, the effects which I have attributed to the Moon's position in declination, are distinct from those before shewn to arise from her change of place in revolution.

The Moon was in her Third or Last quarter on the morning of the first day of 1807 ; she returned to the same phase, after having made twelve revolutions in her orbit, early in the morning of the 22d of the Twelfth month of that year.

There was a New Moon on the afternoon of the 30th of 12th Mo. 1815; and again, after 12 revolutions, on the 18th of the same month 1816.

The reader will find, on comparing these intervals of time with those taken for the declination, that *thirteen* periods of the latter nearly correspond with twelve revolutions ; consequently the Moon must have presented every variety of phase, during these spaces, in conjunction with any given degree of North or South declination ; and every variety of the latter together with any given phase : a state of things which effectually precludes us from ascribing to the one, any variation presented, *upon the whole of a nearly coin-cident space of time,* by the other.

The diminution of the average rain, for the week of Full South declination, was therefore, in 1807, independent of the dryness before ·attributed to the influence of the Full Moon in that year; which was a still more striking phenomenon. Let us see how the case stood, in this respect, in 1816.

Having divided the rain for this year also, according to the phases *about*, and *between* which it fell, and likewise computed the mean Temperature for each of the spaces (which are here denominated weeks) the results are as follow ;

In 1816,

      For the week *about* New Moon,
           Rain 6·11 in.   Temp. 47 10°

      For the week *about* First Quarter,
           Rain 10·10 in.   Temp. 46·60°

      For the week *about* Full Moon,
           Rain 9·13 in.   Temp. 47 17°

      For the week *about* Last Quarter,
           Rain 5·51 in.   Temp. 48·39°
           ————————————————

           Total 30·85 in.   Mean 47·31°

      For the week *after* New Moon,
           Rain 5·21 in.   Temp. 46·85°

      For the week *after* First Quarter,
           Rain 12·49 in.   Temp. 46·88°

      For the week *after* Full Moon,
           Rain 7 41 in.   Temp. 47·78°

      For the week *after* Last Quarter,
           Rain 4·20 in.   Temp. 47·75°
           ————————————————

           Total 29 31   Mean 47 31°

The Full Moon week in 1816, instead of being distinguished for dryness, as in 1807, was excessively wet: the greatest depth of rain, however, fell in the space intervening between First quarter and Full Moon, and the *driest* part of the space included in each Lunar revolution was *in the opposite part of the orbit*, between the Last quarter and New Moon. The reader has only to turn over the Tables, from 114 to 126 inclusive, in the First volume, to be convinced of the fact in each instance.* With the exception of the week following the Summer solstice, in which there fell heavy rain before and after New Moon, the weight of the rain, this year, lies, in a very remarkable manner, within and about the third week of each period, or the space above mentioned; until we come to the latter part of the Eleventh Month and beginning of the Twelfth; when this space suddenly becomes *dry,* and that following the next Last Quarter becomes *wet.* It is observable, though I do not pretend to establish a connexion between the phenomena, that a Solar and a Lunar Eclipse are included in this period, which is so conspicuously dry in this very wet year, the rain being only half the average quantity of the season.

I have remarked that the Lunar orbit, in 1816, appears to have had a wet and a dry side, as it regarded the Moon's influence on the rain of our climate. It appears likewise from the preceding statement, that the Mean Temperature, taken about the phases, was highest for the Last quarter and lowest for the opposite part of the orbit, or First quarter, passing through a mean state for the intermediate phases, of New and Full Moon. Thus the *cold* aspect

* In Table 123, the marks Full M. and 1st Q. have been accidentally transposed.

of our attendant planet was, in this year, also the *wet*
one; and the same arrangement that brought more
warmth, brought also comparative dryness. And this,
as in the very dry year of 1807, subject to a distinct
and independent effect produced by the Moon's declination; to the consideration of which subject we
may now return.

In order to place in a more striking light the effect
of the Moon's *declination* on the *Barometer,* as well
as to shew the agreements and differences in this
respect, of a very dry and a very wet year, I have
given, in Plate 6, four periods of 1807, and as many
of 1816, taken in each case from the winter and
spring; in which seasons these effects are the most
conspicuous. These curves represent the movements
of the Barometer from the day of the Moon's crossing
the Equator, going South, to that of her return in the
same direction to the same position. The regular
curve, which accompanies them in each figure, represents the Moon's course in declination, the horizontal
line being the Equator. In the *upper* figure, the
curves are constructed from the *medium* height of the
Barometer for each day, each of them having its mean
point in the horizontal line. Consequently the reader,
knowing the mean of the period, with the time of its
beginning (both of which are given below), and
availing himself of the help of an Ephemeris, for the
successive times of the extreme South and North
declination, &c. may verify for himself, by the Tables
from 2 to 6 inclusive (vol. 1.), the accuracy of these
delineations; the regular appearance of which, in
some parts, may seem not unreasonably to require
proof by measurement. Yet they are the result of
observations, made without the remotest conception
of their being ever applied to this standard, and in a

manner which I cannot but consider, now, as imperfect.

The *lower*, or second set of curves, give the variations of the Barometer at their full extent, as recorded on the face of the clock, of which I have already given an account in the Introduction to this work (vol. 1. xiii.), and in consequence of their shewing all the smaller variations, which are sunk and lost in the curves constructed from medium heights, their general appearance is very different from the former.

These curves will be found to agree nearly, but not exactly, with the observations in the Tables from 114 to 117 inclusive: the latter having been obtained not from the clock, but from a Barometer in the ordinary way.

In the third or *lowest* figure, the four sets for each year are respectively reduced to a mean curve, which is adapted to a common mean line: and a medium curve, passing between these two, exhibits, finally, the total or average effect of the declination on the Barometer, for the whole of the space taken for this examination.

Time of beginning, and mean height of the Barometer, (represented by the horizontal line), for each of the curves in Plate 6.

For 1807,

Curve	begins		mean line at
a—b	begins	30—31 of 12 Mo. 1806 ;	mean line at 29.97 in.
b—c	...	26—27 of 1 Mo. 1807 ;	... ... 29.76 in.
c—d	..	23 of 2 Mo. 1807 ;	... . 29.92 in.
d—e	...	23 of 3 Mo. 1807 ;	.... 29.79 in.

For 1816,

Curve			mean line at
h—i	...	23—24 of 12 Mo. 1815 ;	. . ... 29.64 in.
i—k	...	19—20 of 1 Mo. 1816 ;	... ... 29.66 in.
k—l	....	16 of 2 Mo. 1816 ;	.... 29.64 in.
l—m	.....	15 of 3 Mo. 1816 ;	... ... 29.79 in.

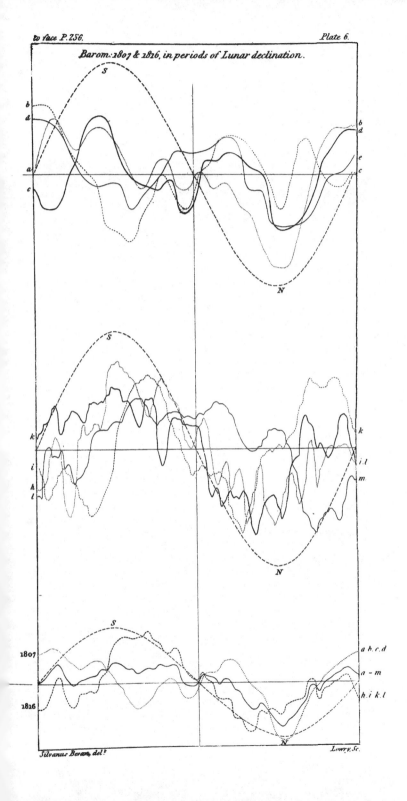

Barom: 1807 & 1816, in periods of Lunar declination.

Silvanus Bevans, del.ᵗ   Lowry, Sc.

For the whole,

Curve *a, b, c, d,* Mean of 4 periods of declination, beginning 30 of 12 Mo. 1806,
ending 19 of 4 Mo. 1807, ... 29.86 in.

Curve *h, i, k, l,* Mean of 4 periods of declination, beginning 23 of 12 Mo. 1815,
ending 11 of 4 Mo. 1816, ... 29.68 in.

Curve *a—m.* Mean of the above 8 periods, ... 29.77 in.

It will be convenient to begin the examination of
these curves with the last, or general one, which, it
will be recollected, gives the daily mean heights of
the Barometer through a period of declination, upon
averages of eight days each; the observations taken
in seasons remote from each other, and under all the
variety of weather to which the winter and spring
months are incident: consequently, in a manner calculated to secure the fairest results.

The general appearance of the curve *a—m* confirms
the position already deduced from calculations on a
larger space. It is, for the most part, above the general mean during the Moon's continuance in South
declination, and below it during her North declination. The depression for the latter is, moreover, the
most regular part of the whole variation; its crisis
coinciding very nearly with the time of the Moon's
beginning to return South, and the times of its de-
parting from and returning to the mean being symmetrical. In this part also, the respective curves of
the dry and the wet year present appearances the
most nearly alike; and it is observable, that in the
dry one the curve descends lowest.

The curve also runs highest in the wet year, on the
South side of the period; when we find the greatest
difference, and indeed opposition, to prevail. While

the Moon proceeds towards the South from the Equa.
tor, the Barometer of the dry year, which had risen
at the going off of the Northerly depression, falls;
and that of the wet year, which had continued, as it
were, struggling below the mean, rises. Two or three
days after the Moon has begun to come back from
the South, each of the curves again changes its direction; that of the wet year now enters on a fall of ten
days, which carries it across the mean to its lowest
point for the whole period: that of the dry year rises
for nearly an equal space, attaining a moderate elevation above the mean; from which it passes into the
Northerly depression. Thus the wet year has the
Barometer at a high level for a week only, while the
Moon is approaching the Equator from the South,
and the remainder of the period may be said to be
nearly occupied by depressions : and the dry year is
subject to a considerable depression, during the week
of Full North declination ; the rest of the period being
chiefly occupied by a mean or elevated Barometer.
And supposing a rule to be found, for the periodical
return of such extreme wet and dry years, we have
here (so far as regards the winter months,) a pretty
certain method of anticipating the time of the occurrence of storms, in the fair season, and of fair and
moderate intervals, in the wet and stormy one. Such
are the mean movements of the Barometer, in these
two seasons so opposite in their character, for the
winter and early part of spring. It was not found
expedient to introduce a greater number of curves
into the figures, or to attempt, in this place, the solution of more complicated appearances. We may now
therefore advert to these curves singly, or as groups,
in order to inquire into the attendant winds, and other
circum tances.

The elevations belonging to the week in which the
Moon was crossing the Equator, southward, in 1807,
constituting the extreme parts of the four curves, were
accompanied by winds from the *South-west, West,* and
North-west. There appears but one observation of
NE, and two of E, in this interval; and not one of a
South wind.

The movements, in 1816, for this space, in which
depressions predominate, had winds from the *South-
east, South-west,* and *West.* Three or four cases only
of a Northerly wind appear, along with the great
elevation in curve *i—k,* continued in *k—l.* This was
at the going off of the severe frost of that season, in
which the Thermometer stood a whole night 5 below
zero. The crisis of the Barometrical depression, on
this occasion, fell on the morning of the 7th of Second
month, which is the date of the lowest point of the
curve *i—k*; and the same winds which brought that
intense cold, produced also the great rise of the Barometer.

For the week of Full South declination we have, in
1807, for the most part *North, North-east,* and *West*
winds: the depression at this time in the curve *b—c*
was effected by South, South-west, and West winds.
In 1816, we have for this space an alternate play of
winds; the *South, South-west,* and *North-east* predominating in *h—i* and *k—l,* and the *North, North-
east, East,* and *South-east* in *i—k* and *l—m,* with
appropriate movements in the curves. The curve *i—k*
exhibits a fine upward sweep of five days under NE
and N winds, after being three days depressed by the
South-east: the crisis of these two movements will be
found in Table 115, at the 24—25 of First Month.
Table 114 will also furnish interesting particulars of
the curious sudden depression, immediately preceding

the great rise which distinguishes this portion of the curve *h—i.*

We come next to the week of mean declination, the Moon going Northward, in which the two movements again cross each other. The winds here are, in 1807, the *South-west, West, North, North-east,* and *North-west,* without any South or South-east; and in 1816, *the South-west, South, West, North-west,* and *South-east*; without any North or North-east, till we come towards the close of the series. Hence the curve *l—m* presents an exception; being kept up for five or six days, where the others fall, by North-east and East winds, and at length falling (out of course and where the others rise) by the progress of the wind to South-east and so round to the Westward. This exception, which followed the Vernal Equinox (see Tab. 117,) extended also to the weather, there being hereabouts seventeen days in succession free from rain, the longest *dry* space in this year!

If we now turn back to the curves for 1807, we shall find in *a—b* a parallel exception. This curve, in crossing the mean line, *descends* on the whole, from the 11th to the 16th of the First month, with a fine movement of decreasing undulation, and with the winds as follow; W, SW, NW, W, SW, NW. It then enters upon the regular depression for North declination: for the particulars attending this and the preceding movement, the reader may consult Table 3 in the First volume. It is probable these movements will be found, hereafter, to be necessary compensations in an extensive system.

There remain now to be considered only the depressions in the week of Full North declination. In 1807 these are very regular, and their crisis agrees nearly with the time of the Moon's being farthest North : in

1816, on the contrary, we see them accelerated or retarded ; so that the crisis, where it can be defined, lies considerably on one side or the other of this point. The difference would have been still more perceptible, had the curves of 1816 been formed, like the other, from *medium* observations.

These depressions are not necessarily attended with gales of wind or heavy rain, at the place of observation. The crisis of that in the curve *b*—*c* was, however, connected in our district with a very severe gale from the NE, with snow and electrical discharges from the clouds ; as that of the curve *a*—*b* probably was, with a storm at a considerable distance, in Devonshire, which appears by the accounts in the papers to have done much damage. See the dates, First Month 21, Second Month 17, Third Month 17, and Fourth Month 13, in the Tables from 3 to 6 inclusive.

In 1816, however, the desultory movements of the Barometer in the lower part of the scale, in this space, did not in many instances baulk the observer's expectation, and there occur in the Tables from 114 to 117 inclusive, all the varieties of foul weather, in connexion with them ; the particulars of which it is not needful here to point out.

With regard to the *direction* of the winds in this space; in the four periods of 1807, the South-west predominates, and next to it are the North-east and North-west, the South-east again absent : but in 1816, the winds are a perfect mixture, there being no point without at least two observations, and the South-west only considerably exceeding in number.

The fairest mode of comparing the winds for these spaces is, however, upon the whole year. I have accordingly taken out the observations for these two years, in spaces answering to those of the Table of

Quarter-periods of declination for 1807, page 246—7; and to those of a similar Table formed for 1816, the results of which are given with the former.

Proportions of the different Classes of Winds, in Quarter-periods of Lunar declination, from the 3d of the 1st Mo. (Jan.) to the 23d of the 12th Mo. (Dec.) 1807; being 355 days, or 13 periods of declination.

	Full South Declination.						Mean Declin. Moon going N.					
Per.	Days	N—E	E—S	S—W	W--N	Var.	Days	N—E	E—S	S—W	W--N	Var.
1	7	2		1	4		7		1	2	4	
2	7	1		2	4		7			4	3	
3	7	5			1	1	7	5			2	
4	7	6				1	7	2		4		1
5	7		2	3	1	1	7	1	4			2
6	7	3	2	1		1	6	1	2	1	2	
7	7				7		6	1			3	2
8	7			4	3		7			4	1	2
9	7	1	2		4		6		5		1	
10	7			1	6		6	4			2	
11	7			3	4		7			5	1	1
12	6	1		3	1	1	7		1	3	3	
13	7	2		1	3	1	6	1		1	4	
	90	21	6	19	38	6	86	15	13	24	26	8

N. B. The spaces taken are, as nearly as possible, those which have the Moon's greatest N or S declination, or her position on the Equator in their middle. The Winds are taken from the Tables in Vol. 1.

Full North Declination.						*Mean Declin.*	Moon going S.				
Days	N—E	E—S	S—W	W--N	Var.	Days	N—E	E—S	S—W	W--N	Var.
7	1	1	2	2	1	6			2	4	
7	1		3		3	6			2	4	
7	3		1	3		7	2	2	1	1	1
7	1		4		2	6	4	1			1
7			4		3	7	1		5		1
7			2	2	3	7			4	2	1
7	2	1	1	2	1	7	2	1	2	1	1
7			6		1	7	1		3	2	1
7		1	3	3		7			1	6	
7	1		4	1	1	7		1	3	2	1
7	1		1	3	2	7	1	3	2	1	
7	3			1	3	7		1	4	2	
7				5	2	7		3	1	2	1
91	13	3	31	22	22	88	11	12	30	27	8

Proportions of the different Classes of Winds, in Quarter-periods of Lunar declination, from the 28th of the 12th Mo. (Dec.) 1815, to the 16th of the same, 1816; being 355 days, or 13 periods of declination.

	Full South Declination.						*Mean Declin.* Moon going N.					
Per.	Days	N—E	E—S	S—W	W—N	Var.	Days	N—E	E—S	S—W	W—N	Var.
1	7			5	2		7			4	3	
2	7	5	2				7		2	5		
3	7	1		4	2		7			4	3	
4	7	4	1		2		7	2	5			
5	7	1	2	1	3		7	3	4			
6	7	2	1	1	3		7	4	1	2		
7	7	3	1	2	1		7	3		2	2	
8	7			4	2	1	7		1	3	1	2
9	7			5	1	1	7		2	3	2	
10	6		2	1	3		7			4	3	
11	7	1		4	1	1	7		4	1	1	1
12	7		5	1		1	7	1	2	3		1
13	6	1	3	1	1		7	3		2	2	
	89	18	17	29	21	4	91	16	21	33	17	4

N. B. The spaces taken are, as nearly as possible, those which have the Moon's greatest N or S declination, or her position on the Equator in their middle. The Winds are taken from the Tables in Vol. 1.

Full North Declination.						*Mean Declin.* Moon going S.					
Days	N—E	E—S	S—W	W--N	Var.	Days	N—E	E—S	S—W	W--N	Var.
6		1	3	2		7		3	3	1	
7	4	1	1		1	6	1		3	2	
7	2		4	1		7			4	3	
7	1	3	1	1	1	6	3	1		2	
7		2	1	4		6			4	2	
7	2	2	1	2		7			1	5	1
6	2		1	2	1	7			1	2	4
6			4	1	1	7	1			5	1
7	2			4	1	7	3	2		2	
7	2	1	4			7	3	3	1		
6	1	2	1	1	1	7		1		4	2
7	1		2	2	2	7				3	4
7		3	3	1		7			2	3	2
87	17	15	26	21	8	88	11	10	22	35	10

Summary of the distribution of the Winds according to the Moon's declination in 1807 and 1816.

1. With the Moon full South.

	N—E	E—S	S—W	W—N	Var.	Days.
1807	21	6	19	38	6	90
1816	18	17	29	21	4	89

2. With the Moon coming North.

	N—E	E—S	S—W	W—N	Var.	Days.
1807	15	13	24	26	8	86
1816	16	21	33	17	4	91

3. With the Moon full North.

	N—E	E—S	S—W	W—N	Var.	Days.
1807	13	3	31	22	22	91
1816	17	15	26	21	8	87

4. With the Moon going South.

	N—E	E—S	S—W	W—N	Var.	Days.
1807	11	12	30	27	8	88
1816	11	10	22	35	10	88

Totals		N—E	E—S	S—W	W—N	Var.	Days.
	1807	60	34	104	113	44	355
	1816	62	63	110	94	26	355

The two classes N—E and S—W are of nearly the same *total amount* in the wet, as in the dry year. The character of a whole year, in this respect, does not appear to be decided by either of them; but rather by the class E—S, which has nearly twice the amount in the wet year, that it exhibits in the dry: and this excess is taken out of the class W—N, and out of the *variable*. In regarding the year as a whole, it is also proper to remark, that a much greater quantity of air undoubtedly passed over us, in all directions, in 1816 than in 1807 The large amount of variable winds, which appears under the Full North declination for

1807, is clearly raised at the expence of the E—S and N—E classes. I am not conscious of having used less care respecting these classes in 1816; and am inclined to believe that, during the fine season of 1807, there prevailed a much larger proportion of variable *Easterly breezes,* than of *winds* from either of these quarters. It seems to be one of the conclusions of such a season, that the air of the district shall not hastily travel out of it, nor that of a distant one suddenly invade it. A windy season can hardly fail, at least in some part of it, to be a wet one.

The distribution, as well as the amount, of the N—E is nearly alike in the two years. I shall therefore leave it for the present, to attend to the next in order.

The class E—S, which I have already characterised as the principal *importer* of our rains in Spring and Summer, appears to make its way into this district chiefly while the Moon is approaching from the South. The air being thus vapourized to the degree required for the moderate rains of the season, this wind falls off, in the dry year, during the week of North declination, to a very inconsiderable quantity: but in the wet year, it is reduced to its minimum, only during the return of the Moon to the South.

The class S—W follows nearly the same rule. It increases as the Moon comes North, and decreases as she proceeds South again: but it is more fully manifested, under Northerly declination, in the dry, than in the wet year; continuing nearly undiminished until the Moon is Full South.

Northerly winds are of course more frequent in those seasons when the Southerly fall off. They were at their height, in 1807, in both classes, under Full South declination: the W—N, in this year, came to their minimum under North declination, the N—E not until the following week, when they were only at about half their greatest amount.   In 1816, the class W—N appears to have supplied the place of the N—E, while the Moon was going South; falling to half the number in the week of her return Northward, and exhibiting a mean amount in the intermediate weeks.

This account of the Winds, compared with the summary of the effects on the Barometer, Temperature, and Rain, in page 251, may supply us with a key to many of the facts there stated.

A general tendency in the Northern atmosphere to come over us, while the Moon is far South, may be admitted as a cause why the Barometer at this time is above the mean, the Temperature about or below it, and the Rains in small quantity.

As the Moon comes North again, the air returning from the South causes increased temperature : it brings also a great increase of vapour, and the heat evolved during the condensation of this, may possibly be the means of the greater elevation of the mean Temperature at this time, in the wet, than in the dry year. Something must however be attributed, in this case, to the actual translation of more of the tropical air into these latitudes, in a wet season. The increase of the rain at this time, in both seasons, is a necessary consequence of the other arrangements.

Why the Barometer should now be below the mean in the dry, and at its highest average in the wet season, is not equally apparent: but we may further notice its movements in the conclusion.

Under Full North declination, we have the results of the previous introduction of vapour by Southerly winds. In the dry year, the vapour is decomposed in a short space of time, and the attendant gales of

wind are single and decided : in the wet, a longer
continuance, or a number of repetitions of this process, together with the larger product of rain, indicate
the operation of numerous currents from distant regions. In each season, these causes suffice to bring
the Barometer to its lowest average, and the Temperature to the mean.

While the Moon is returning to the South, the winds
from West to North predominating, in the wet year,
tend to raise the Barometer and reduce the Temperature. The latter effect may also be now accelerated
by Evaporation, as the rains decrease again. It is
remarkable that, in 1807, the Barometer shews the
highest average for this week, and the Temperature
the lowest, with the smallest proportion of N—E, and
nearly the largest of S—W winds.

The course of the varying density of the atmosphere in its relation to the Moon's declination, is
pretty fairly represented as to direction, though not
as to extent, for the whole of the two years, by the
specimen given in the two mean curves, *a, b, c, d,*
and *h, i, k, l,* in Plate 6. It will be important, hereafter, to ascertain fully the principles of these two
modes of variation ; as they appear, more than any
other circumstance (the disproportion of the South-
east winds excepted) to mark the difference between
a wet and a dry season ; and their periodical causes
being once known, the return of such seasons may
be predicted with some degree of certainty. I consider the scheme which I have given early in this
inquiry (page 94) of the varying *mean Temperature*
of the years, as calculated in great measure to answer
this purpose ; it being very clear, that the greatest
depth of rain falls in the coldest years, and that the
warm years are dry or mean ones. But it will be a

great addition to this information, should we be able to prove, from observations now extant, that the Barometer also varies its mean height periodically, from year to year; and that both variations are governed by a periodical succession of the different classes of winds.

———— ————

I might add to the mass of evidence on this subject some proofs of a peculiar relation between the Moon's *apogee* and *perigee,* and the mean height of the Barometer on the days on which they occur: but I have nothing as yet, so far digested as to be relied on. Indeed the labour of preparing what has now been thrown before the reader, has greatly exceeded my expectations; and being prosecuted with considerable disadvantage, in the midst of other engagements, has delayed rather unreasonably the completion of this part of the work. In publishing, in their present state, so large a proportion of the facts derivable from my observations, I shall undoubtedly throw the whole remainder of them open to the use of others, and may probably be thus anticipated in some important deductions yet to come. But I am not at all jealous of the little merit which attaches to discovery, in a field so rich, and hitherto so little trodden; and shall be well satisfied should others, to whose minds the requisite knowledge of Astronomical relations may be familiar, and their capacity for such enquiries, from a mathematical education, greatly superiour to mine, be willing, after examining these data, and correcting such errors as they may find, to take up the subject, and improve upon my beginnings.

If the Moon's attraction be really the principal cause of those variations in the atmosphere which

cannot be traced to the influence of the superiour
planet, the mode of operation of this attraction may
be very simple, at the same time that, considering the
complicated nature of the Lunar orbit, and the perpetual interference of the Sun's varying power, its
manifestations in any given temperate climate may
prove a very difficult subject to investigate.

On a train of effects, the most part of which are out
of the reach of direct observation, we may be permitted, in this part of the work, to hazard a few conjectures.

The surface of the atmosphere is, I think, less elevated, and better defined, than many persons would
be led to imagine it. A portion of air, rarefied by
means of the air-pump, does indeed exhibit an elasticity, which seems limited only by the imperfection
of the instrument. For the most minute residuum
still appears to fill the vessel, and to press against it
in all directions. But it does this at a temperature
which, compared with that of the extreme boundaries
of the atmosphere, is probably that of the steam in a
high-pressure engine to the water in a well. We know
that, in ascending in the atmosphere, the temperature
is found to decrease with the decreasing density of
the air: and even under a vertical sun, between the
Tropics, a line of perpetual snow on the mountains,
indicates a boundary within our reach, which the heat
never has ascended in mass to penetrate. There is
consequently no source, from whence air conveyed to
the summit of the atmosphere, could take the heat
necessary to such extreme rarefaction : the whole sensible heat of the atmosphere being derived originally
from the earth's surface, and distributed in an inverse
proportion to the elevation. At an elevation, therefore, not perhaps on a mean more than ten times that

of the highest mountains, or fifty miles at the Equator, and considerably less at the poles, I conceive there exists a perpetual *zero* of temperature; and with it an effectual limit to the further expansion of the atmosphere. Here, the spheroidal body of gases, enveloping our globe, has probably a well defined surface (its extent considered) where the air, though greatly attenuated, is much less rare than we can make it in the receiver of the air-pump; in a word, a fluid, capable of rising and falling, like the waters, by change of gravity.

With such a surface, it is plainly possible that the atmospheric ocean may be acted on in the manner of a tide. It may be elevated and rarefied on the side directly opposed to the Moon, and at the same time on the opposite side of the globe; and left to its proper gravity in the remaining part of the mass. And it ought, on this supposition, to exhibit a more perfect example of a tide than even the waters; there being here no shores, as in the ocean, to retard the arrival of the swell at a given place, at the destined hour; or prevent its passing regularly round the middle regions of the globe, in the space of a revolution of the latter on its axis. If I place my hand upon a spiral spring of wire, and depress it, the force being withdrawn the spring follows, and returns immediately to its former state. But if I do the same with a pillow of down, this elastic body, consisting of many small parts acting feebly on each other, takes a long time to resume its full dimensions. There is a similar difference in constitution between the ocean and the atmosphere: and it is very probable, that an interval of six hours is not nearly sufficient for the full effect of rarefaction, and still less for the subsidence and condensation of the air, through its whole depth, to

the degree required by the theory of such a tide.
The *daily* alternate movements, then, of an atmospheric tide, perhaps from their not having been sufficiently sought among the continual fluctuations of
the density of the air at the earth's surface, are not
yet demonstrated : but both the Barometer and Thermometer supply, in their respective mean variations,
most palpable instances of the weekly increase and
decrease of those movements.

In a portion of the atmosphere, the most considerable in point of bulk, situated above the reach of the
daily variations of temperature caused by the sun, the
alternate rarefaction and condensation here supposed
may take place, without producing any other consequence than a current from East to West, around
the globe, in that region.

In a lower portion, visited at different times by different proportions of the heat and vapour generated
at the earth's surface, it may effect an alternate absorption and condensation of water, with correspondent
changes in the electrical state of this region ; and thus
contribute to decide the occurrence of strong winds,
rain, thunder, and other occasional meteors, below.

Still lower, in a region to which our observation
more or less extends, the complexity of the causes
must necessarily produce effects more difficult to
appreciate ; and these are brought about, as it seems,
chiefly by the succession and interchange of lateral
currents. The rarefaction produced in this region by
the Sun's heat, is admitted to give rise to a most
regular and extensive system of these, commonly
called the *Trade-winds.* The air, around the globe
over the Equatorial regions, expanded by the heat
rises, or is pushed upward by the contiguous cooler
air from the North and South ; the motion of which

combined with the larger motion of the earth's surface
from East to West, as the latitude becomes higher,
gives rise to a South-east wind on the South side, and
a North-east on the North. Such is the admitted
principle of the Trade-winds; and it is thought (indeed it must follow) that the air, thus elevated above
the Equator, returns in some kind of currents, above
the Trade-winds, towards either pole. If we admit a
constant Easterly *tide* in the higher tropical atmosphere, into which this rarefied air constantly rises, we
shall have a powerful auxiliary to the Sun, in keeping
up the Trade-winds; and if we admit that the Moon,
by her alternate passage to the North and South in
declination, sets this tide alternately to the Northward
and Southward of the line, we shall have a principle,
on which to solve the greater abundance of rain, and
brisker flow of the variable winds, in temperate latitudes, at that season when the Moon becomes vertical
to them, than when she is in the other hemisphere.
We want indeed, on this point, the concurrent testimony of observations made in some temperate climate, South of the Equator: though we know already
that their *polar* winds raise the Barometer, while the
Equatorial depress it; following the same law as with
us, though moving in opposite directions.

The air which flows from the North and South
towards the Equator, is felt as an Easterly wind, because it comes from parts of the earth's surface,
which have a smaller motion from West to East, than
the region into which it is entering : but it gradually
acquires the rotary velocity proper to that region. In
returning to the temperate latitudes, it has again to lose
this Westerly momentum : and this seems to be the
principal cause of the great preponderance of Westerly
winds in our own climate. We have seen, page 157,

that on a mean of ten years, the Westerly were to the Easterly winds as 225 to 140; while the Northerly and Southerly winds balanced each other within 21 days. A wind, coming to us from a considerable distance South, whatever be its velocity, must therefore be felt as a South-west wind : and as the Trade-winds, at certain seasons, appear to have their subsidiary streams, or appendages reaching far into the colder latitudes, so these Tropical Southerly gales occasionally make their inroads upon us with greater violence, and for a longer season than usual. It is not unlikely that the British isles, in consequence of their latitude, and from their being as it were a part of its Western barrier, may be the very part of Europe the most exposed to them. A North-east wind, kept up by rarefaction caused by the sun, must find the easiest course upon land ; while a South-west, consisting of air which has to descend upon the earth, and spend the momentum it has acquired in more Southern latitudes, is more likely to get easily over the surface of the ocean, and to be arrested by the asperities of the first extensive fixed surface which it encounters. This state of things prevailed remarkably near the close of the winter of 1817, after a long course of violent Westerly gales; when, as far as we may judge from the reports of navigators, the North-east current was for many days no further to the South of us than the coast of Portugal, without our feeling even the skirts of it. See the note at the bottom of page 7 in this volume, and the Results of the next Table.

It is remarkable nevertheless, that on a mean of ten years, ending with 1816, and indeed in most of these, taken separately, the winds to the North and South of West should so nearly balance each other, as that their averages stand 100 : 104. I know of no reason

which can be given for this, equally satisfactory, with
that of their receiving a direction to North and South
alternately, by the Moon's different positions in declination : an effect which, although not to be found
in the winds of a particular district, in each Lunar
revolution, is yet detected in a long average.

On the whole, it may be inferred that the winds in
a temperate latitude, like our own, after escaping
from the Tropical vortex, become subject, in winter
more especially, to the Moon ; and that their tendency
is, to follow her path, or the moveable point of greater
rarefaction which she marks out for them.   Thus it
appears from the statement, p. 266, that during her
approach to these latitudes, in declination, in 1807
and 1816, the winds from the East and South-east increased, while those from the West and North-west
fell off in number. Now if we consider that the Moon's
daily course from East to West (which though only
apparent has here the same effect as the real,) was
coupled, during these weeks, with a motion from South
to North in declination, it will appear that a South-
east wind would now, in effect, follow her course, and
a North-west flow in opposition to it.   And in 1816,
during the weeks in which the Moon was receding, in
declination, to the Southward, and thus offering daily
less and less resistance to a North-west wind, this class
of winds amounted to double the number which they
exhibited in the former case.   Again, in both years,
and especially in 1807, the class of winds from North
to East, which are plainly most influenced by the Sun,
appeared in the greatest number while the Moon was
in Full South declination, and when consequently
there was little of the rarefaction, which she is here
supposed to produce, in these latitudes.

The succession and proportion of the winds are

consequently subject to a periodical variation from
year to year : but the period in which the same, or a
similar set of winds comes round again, cannot at
present even be conjectured. From the effects produced, in our district, on the average temperature of
the years, and on the depth of rain, it may seem to
have some connexion with the Lunar cycle of 18 years.
But this is a subject well worthy of separate and more
deliberate investigation. In what I have brought
forward, I consider myself to have redeemed the
pledge given in the introduction to this work. (vol. 1.
page v.) I think I have decided in the affirmative (as
it regards the phenomena of our own district,) the
first part of the question there proposed, " *whether*
the relative positions of the Moon, in the different
parts of her complex orbit, influence the state of our
atmosphere." I have also thrown " some light" on
the second part, which proposes to inquire " in what
way" this effect is produced. Should it prove so
much as shall suffice to stimulate the ambition of
Astronomers, in different parts of the world, to annex to the stupendous field of their present labours
this lower (if they please so to consider it) and
almost uncultivated province, my purpose will be
answered ; and I have no doubt the consequences
will be beneficial to mankind. For although it be a
very just remark, that the seasons would not only not
go on better for our purposes, but would be in utter
confusion, had mankind the *ordering* of them ; it does
not thence follow that, could we calculate their periods
and foresee their extremes, both our personal safety
and comfort, and the success of our labours, might
not be essentially promoted by such foreknowledge.

An ample, extensive and accurate collection of facts
for each climate, is therefore the first *desideratum,*

These should be digested by each observer from his own observations, or from those made in his own district at least, where his local knowledge may greatly facilitate the work. They may be cast either into the forms I have here adopted, or into such others as may be preferred; but as much as possible in a way to be comparable with the results of others. The materials will be found more ample than many would suppose them. To give a single prominent instance, the "Meteorological Observations made at the Royal Observatory at Paris" contain a mine of treasure which it would require years of labour for any single person to explore, in the way in which I have gone through my own and the adjunct observations, belonging to our London district. And I have no reason to think, from the appearance of the few parts I have examined, of those belonging to Paris, that their results would be less regular and systematic than those contained in this volume.

It will be necessary, before this section be dismissed, to give some account of the general Tables, F and G, at the end, entitled " Mean Results of Lunar periods arranged by the Solar year." Wishing to place before the eye collectively, in some part of the work, the results of the Barometer and Thermometer, for the Lunar periods in which my observations had been published, I cast them into the form there exhibited, putting, as nearly as possible, all those which comprised the Solstitial or Equinoctial points, under each other in the same column, and throwing a few periods into a kind of intercalary space at the ends, for this purpose. The date and extent of

each of these periods may be found at once, by referring to the Table under the number. They vary in
each column, as to date, through a space of not less
than 25, nor more than 29 days : consequently each
column ranges through a mean space of 55 days :
the intercalary results, however, which are cast, in
the following averages, along with the first and last
columns, add somewhat to the extent of those columns.

The only use which I shall make of these Tables
at present is, to give the results of the first ten years
in quarterly averages, and deduce some consequences
from them. The reader will however notice the regular gradations which the averages at the foot of each
Table present ; on which subject, as it respects the
several *months* of the year, I have already treated.

> *Averages of Temperature for 38 Lunar Revolutions,*
> *beginning at New Moon, and for 86, beginning at*
> *Last Quarter, the whole comprehended in a space*
> *of 10 Solar years; from the 10th of Twelfth mo.*
> Dec. 1806, to the 11th of the same, 1816.

1. *Brumal Periods.* Average of ten periods
   in the second column and two intercalary   37·92°
   Of ten in third column .   . . . .   35·73
   Of ten in fourth column   . . . .   39·63

   Of thirty-two periods   . . . . . .   37·76

   Below the Autumnal 11·61°

2. *Vernal Periods.*
   Average of ten in fifth column   . .   42·25
       of ten in sixth column   . .   48·92
       of ten in seventh column   . .   55·67

       of thirty periods   . . . . .   48·94
   Above the Brumal 11 18°

3. *Estival Periods.*

Average of ten in eighth column . . . 58·62•
　　　of ten in ninth column . . . 62·39
　　　of ten in tenth column . . . 60·99

　　　of thirty periods . . . . . 60·66

Above the vernal 11·72°

4. *Autumnal Periods.*

Average of ten in eleventh column . 56·70
　　　of ten in twelfth column . . 50·75
　　　of ten in thirteenth and two⎱
　　　　　intercalary . . . . ⎰ 40·68

　　　of thirty-two periods . . 49·37

Below the Estival 11·29•

Averages of the Barometer for 124 Lunar periods,
beginning and ending as before stated respecting
Temperature.

1. *Brumal Periods.*

Average of ten periods in the second
　　　column and two intercalary 29·745 in.
　　　of ten periods in third column 29·788
　　　of ten periods in fourth column 29·874

　　　of thirty-two periods · . . 29·802

Above the Autumnal ·021 in.

2. *Vernal periods.*

Average of ten periods in fifth column 29·870
　　　of ten periods in sixth column 29·814
　　　of ten periods in seventh column 29·812

　　　of thirty periods . . . . . 29·832

Above the Brumal ·030 in.

3. *Estival periods.*

Average of ten periods in eighth column 29·899 in.
    of ten periods in ninth column 29·879
    of ten periods in tenth column 29·854

  of thirty periods . . . . 29·877

Above the Vernal ·045 in.

4. *Autumnal periods.*

Average of ten periods in eleventh column 29·883
  of ten periods in twelfth column 29·736
  of ten periods in thirteenth } 29·725
   column and two intercalary }

  of thirty-two periods . . 29·781
Below the Estival 0.96 in.

I consider that by this mode of averaging the *Temperature*, the inequalities, or deviations from the mean of the season, which I suppose to be produced by the Moon's power over the winds, and which are sufficiently apparent (to the extent, indeed, of ten or twelve degrees in most of the columns) in Table G, are completely done away; and the Temperature restored to the course which it ought to have, by the action of the Sun's power alone. The four quarters, accordingly, rise and fall in nearly equal progression; each being, on a mean 11·45° warmer or colder than the preceding quarter.

I consider that, by the same method, the Lunar influence on the *Barometer* is also done away, and the averages of this instrument brought, in Table F, to the state in which they would be found, in each season, had the Moon nothing to do with them. Any remaining inequalities may therefore be fairly attributed

to the temperature, and to what may be termed the
Solar succession of the different classes of winds
through the year; which is exhibited, as to the
calendar months, over the diagram of the rain, in
page 193.

Under these circumstances, while the Temperature
of the several quarters rises and falls in regular progression, the inequalities of the Barometrical heights
follow a very different rule. The winter Barometer
gains, in its average, ·021 in. upon the Autumnal;
the Vernal ·030 in. or half as much more, upon the
Winter; the Summer ·045 in. or half as much more
still, upon the Vernal; but in the Autumnal average,
the whole difference is lost again, and the Barometer
comes back to its lowest level.

Now, with regard to the seasons in which the
Barometer stands highest and lowest, much may be
attributed to the reigning winds.

Thus, the first Estival result, which is the highest
of the whole series, lies in the midst of the W—N
winds; and the two latter Autumnal ones, in which
the mean is depressed to its lowest point, come after
a long course of predominant S—W winds. Again
the N—E class may be thought gradually to elevate
the Brumal periods, and keep up that in which the
Vernal Equinox is included; while a subsequent
mixture of Southerly winds, in the spring, gives
occasion to some depression before the return of the
high mean about the Solstice.

But there is a probable cause for this gradation
which must not be overlooked, and which has in fact
an equal claim with the winds to consideration. The
mean state of the Barometer in any moderate district,
it is well known, does not represent the weight of the
air in that district alone, but for a great extent around

it : in which extent different winds may even be found
to predominate through the same period of time.
And no reason can be given, more appropriate, why
the Barometer should rise under a certain course of
winds, than that the atmosphere is then receiving an
addition to its ponderable mass; or why it should
fall, under another course, than that it is then sustaining a loss in this respect. The loss and the gain
consist in water; which is at one time converted into
vapour, permanent as a part of the atmosphere for
the season, at another dismissed in rain. Now, in
the Brumal quarter, where we find the average of the
Barometer lowest, the Temperature is lowest also ;
and there is every reason to conclude that the atmosphere in our district, and for many degrees of latitude
and longitude around us, contains, at this season, the
lowest proportion of ponderable vapour. As the
spring comes on, in these latitudes, and the air
acquires heat upwards, it acquires also vapour, and
therefore weighs more on a mean than in winter.
In the summer months, yet more heat and more vapour are accumulated, and the weight of the whole
atmosphere attains its maximum. The addition in
each of these seasons is in a greater proportion than
that of the heat; probably because the higher the
latter ascends, and the more rare the medium is, in
which the vapour is diffused, the greater the quantity
which an equal addition of heat can maintain in its
elastic form. At length comes the Autumn, in the
course of which the Sun retires to the Southward,
the atmosphere of these latitudes cools and collapses
throughout, a great proportion of the vapour it held
is decomposed, and its water deposited in extensive
heavy rains; and the air, losing this portion of its
mass, returns to the former low standard of gravity.

Such are the considerations which it seemed needful to take into view, along with the succession of the winds, in accounting for this gradation in the mean height of the Barometer. Should they be founded in fact, a similar gradation will be discovered, by using similar averages, in correct Registers of the Barometer for nearly every part of Europe.

It may perhaps be worth while to include, in any future researches into the variations of temperature connected with the Moon's positions, the question, whether there exists any communication of heat between the two planets *by radiation.* It is a received opinion, but I doubt whether founded on any experiments sufficiently accurate or delicate, that the rays of light which we derive, by reflection, from the Moon, bring no portion of heat whatsoever along with them.

The two planets are certainly very differently circumstanced as to temperature. The Moon being so much the smaller body, and presenting in consequence a much more convex face to the Sun, would, if it were acted upon in the same manner by the Sun's rays, derive from them less heat, and possess in consequence a lower mean temperature than the Earth. But the surface of this planet is presented after a very different manner to the influence of the solar rays. From the time that they impinge on any given part in longitude to the time of their quitting it, a period elapses, equal to twenty-nine of our mean daily periods of sunshine. And the same part, having once emerged from the rays, has an equally long space allowed it to cool again, in uninterrupted darkness.

Moreover the rays, which thus act through a day of
two weeks duration, are not as in the case of our
polar regions, very obliquely received; but fall on a
considerable portion of the surface more or less directly. The effect of this arrangement must be, that
the middle regions of the Moon, at least, would experience the extremes of heat and cold, in a way to
which no part of the Earth's surface can afford a
parallel—unless the vicinity of the latter should prove,
by reciprocal radiation, the means of equalising the
temperature, in some degree, in both planets.

I mean simply to state it as a possible case, that
the Full Moon, with a surface intensely heated by
the Sun, may radiate a portion of heat to the colder
parts of the Earth's surface towards the poles; more
especially when in her extreme North and South
declination: and, on the contrary, that the New
Moon, having become proportionately cold on the
surface opposed to us, may receive by radiation from
the Earth, and more especially from the Tropical
regions, a compensating degree of heat; which may
serve to moderate the rigour of the nocturnal cold on
that planet. These are the extreme cases: but if we
admit the principle, there will ensue various modifications of the effects, according to the different relative positions of the two planets, and of both with
respect to the Sun.

It would be premature, while only two years of
observations in our own district have been examined,
to attempt to apply this theory to the facts. There is
however, something so remarkable in the regularity
of the increase and decrease of the mean temperature
according to the different Lunar positions, in 1807
and 1816, as stated page 241 and 253, that it will
certainly be desirable to examine, whether any thing

parallel to it exists in other climates, more especially
in the Tropical regions; as well as what aspect the
remaining years of the series present, in this respect,
in our own.

To ascertain, in a more satisfactory manner, whether or no there exists any radiation between the two
planets; I would propose that trial be made with
concave metallic mirrors, having the bulb of a very
sensible Thermometer in the focus; in the manner in
which several very instructive experiments have been
conducted, on radiation among terrestrial bodies.   It
is not at all likely that *glass lenses* should detect so
delicate an effect as the one in question.   If the principle here supposed be real, the rays of the Full
Moon, received in the direction of the axis, should
raise the temperature in the focus of the mirror some
degrees, in a high Northern latitude; and depress it,
in situations near the Equator : due attention being
paid in both cases, to insulate the Thermometer, and
secure as much as possible a stationary temperature
in the surrounding medium.

A curious phenomenon, resulting from the play of
light between the two planets, is so obvious to common notice, that I am surprized not to have met with
any sufficient explanation of it.   In the interval between the New Moon and First Quarter, when the
Moon is seen in the Western sky after sunset, the
dark part of the disk, between the cusps and all round
the hemisphere, is sometimes so far enlightened as
to be not only visible but conspicuous.   And in an
equally clear sky at other times, this portion of the
disk in the same situation is not to be discovered.
It appears that the sunshine on our planet is first
reflected to the dark part of the Moon, and from
thence back to the eye of the spectator : and the rea-

son why this effect is at times (and only at times) sufficient to render the whole disk visible to us, may be, *that there is then an extensive surface of snow on the Northern American continent.* It will be found, on trial with a globe, that while *we* are contemplating the Moon in the position abovementioned, the Sun is yet sufficiently elevated over those parts of the world, for the snows to reflect its beams very copiously to the Moon's surface. Some observations, made since this appearance first began to attract my notice, compared with accounts of the fall of snow in America, have given the matter sufficient importance in my view, to induce me to throw it out as a conjecture.— But it is now time to quit these speculations, and take, in conclusion, a summary view of the whole subject of the work.

SUMMARY.

LONDON, or the metropolis of the British empire, collectively so denominated, is situate towards the western extremity of the plain, or valley, forming the estuary of the Thames. The course of this river is on the whole from West to East, through the city; a little below which, a smaller plain opens to the North, watered by the river Lea, which here falls into the Thames. The sea is distant fifty miles towards the South, with pretty high land between; and about as much towards the East, where the Thames joins it. The site thus described is bounded, except in the direction of the estuary, by rising ground, and by hills of moderate elevation, from which other streams descend into the Thames on each side. The soil is loam and gravel, on a substratum of clay: and the drainage and embankment being perfect, the country though in some parts considerably wooded, and in others below the level of high water mark in the river, is dry and healthy

The latitude being 51° 31′ N. we enjoy the Sun in the shortest days, for seven hours and three quarters, and in the longest, during sixteen hours and a half.

The *Mean Temperature* of the *Climate*, under these circumstances, is strictly about 48.50° Fahr.: but in the denser parts of the metropolis, the heat is raised, by the effect of the population and fires, to 50.50°; and it must be proportionately affected in the suburban parts. The excess of the Temperature of the city varies through the year, being least in spring, and greatest in winter; and it belongs, in strictness, to the *nights*; which average three degrees and seven tenths warmer than in the country; while the heat of the day, owing without doubt to the interception of a portion of the solar rays by a constant veil of smoke, falls, on a mean of years, about a third of a degree short of that in the open plain.

The Mean Temperature *of the year* is found to vary in different years, to the extent of full *four and a half degrees*; and this variation is periodical. The extent of the periods, for want of a sufficient number of years of accurate observations, cannot at present be fully determined; but they have the appearance of being completed in *seventeen* years. We may consider one of these cycles, as commencing either with 1790 or 1800, and ending with 1806 or with 1816. See the figure and explanation, p. 94—95.

In either case, a year of mean temperature begins the cycle; in which the *coldest* year falls at the end of ten years, and the *warmest* at the end of seven years, reckoning from the coldest; and thus alternately; both together including a complete revolution of the mean temperature from its higher to its lower extreme—(or *vice versa* from the lower to the higher) and back again. The year 1816, which was the coldest of a cycle, appears to have had its parallels in 1799 and 1782; and there is every reason to conclude, from present appearances, that the warm temperature of 1806 will re-appear in 1823; which will probably be the warmest, and 1833 the coldest, *upon the whole year*, of a cycle of seventeen years, beginning with 1807.

These extreme annual temperatures are gradually produced, and chiefly by elevations and depressions of less extent, which take place in alternate years: and as the character of the year, in other respects, follows the mean temperature, it is very desirable to ascertain, whether similar periods of variation exist in the annual temperature of other European districts, not too remote in latitude and elevation from our own.

The greatest *heat* to which our climate is subject, in the course of one of these periods, is *ninety-six* degrees:—the greatest *cold five* degrees *below zero.* Thus the full range of our Temperature is about 100° of Fahrenheit.*

* In my first volume (Introd. p. xxxiii) I adverted to the desirableness of our " adopting, by consent, uniform modes, terms, and measures," of observation, in order to render more easy the communication between meteorologists in different countries, and thus advance the science. This hint has been taken up by the editor of the Journal de Physique, in a passage of which the following is the substance:— " There continue to be published in Thomson's Annals, the monthly meteorological observations made at Bushy Heath, near Stanmore, by Col. Beaufoy, and at London by Mr. Howard, to whom the science is indebted for the new nomenclature of Clouds, which is generally adopted in England, and already in use in some parts of Germany, though nearly unknown in France."
" Tilloch's Philosophical Magazine, Schweiger's Journal, and Gilbert's Annals have likewise produced observations of this kind, of which, however interesting, we cannot, for want of room, give an account: we shall therefore content ourselves with observing how important it is, that meteorologists should be explicit in their account of the manner and time of their observations, and that their instruments should be comparable with each other. Indeed it would be worth while, in order to secure this point, that a sort of congress of observers should be held, as M. Pictet proposes, to deliberate on the subject. Unless indeed there should appear some elementary treatise, including all the branches of meteorology, and exhibiting a model, sufficiently well executed to overcome the prepossessions, not of individuals only,

Our temperature scarcely ever rises above 80° but the occurrence is followed, either in our own or in some neighbouring district, by a thunderstorm. These tempests are apt to be more severe, and of longer continuance, in our plains, than in the more hilly or mountainous districts of the island; the equilibrium between the clouds and the earth being, here, less easily restored. They appear to be a consequence of the irruption, upon our previously calm atmosphere, of the temperate air of the Atlantic; they are followed by more or less of rain, and by a reduction of the heat for a season.

With regard to the other extreme, we are so situated, that even in the coldest season of the year, the medium of the twenty-four hours, *upon a long average*

but of learned societies and nations; and thus secure for itself universal reception. If we are to expect such a treatise at all, it appears that it must be from the pen of Mr. Howard, of whom we have just made mention; and who, in effect, has published, in the course of this year, the first volume of a work, entitled the *Climate of London,* which appears likely to fulfil the object." Janvier 1819, p. 31.

However willing I may be, to contribute what lies in my power, to a general good understanding among the cultivators of this science, I must here, once for all, disclaim in favour of some more qualified leader, the pretensions above described. In the mean time, I may be permitted to advance a modest plea in favour of Fahrenheit's scale of the Thermometer, at present used by British meteorologists. There is a convenience in its extent, and even in its mode of graduation, which I should be loath to resign in favour of one, the divisions of which should be either so large, as to require a resort to fractions in every observation; or so minute as to burden the memory, and make it difficult to seize and retain its prominent points and relations. And similar reasons induce me to prefer our own graduation of the Barometer to that at present in use at Paris. Yet there is no doubt with me, that our observers would be disposed to sacrifice, in some degree, their convenience and their predilections, for the great object of a common uniform standard.

of years, does not fall below the freezing point. Continued frost in winter is, consequently, always an exception to the general rule of the climate. The winter even passes, occasionally, almost without frost : in return for which, we have, at uncertain intervals, a rigorous season of many weeks duration, attended with the deep snows, and clear atmosphere, common to more northern latitudes. Our seasons of frost go off, like those of great heat, with a wind from the Atlantic.

The greatest heat falls, on a mean of years, not about the summer solstice, but at an interval of a month after it, and the greatest cold, at the same interval after the winter solstice. The mean temperature of the year is, in like manner, developed at an interval of about a month after either equinox. The nature and reason of this curious law, together with the daily gradation of the heat through the year, which I have fully treated under the head *Temperature,* does not admit of recapitulation here.＊

One of the most remarkable features of our climate is, certainly, the great *variableness* of the temperature : which departs from the mean in either direction, in the course of a few days, and sometimes in a single day or night, to an extent greatly exceeding that which the simple presence or absence of the sun would, at that season, occasion. This appears to be the basis of the so much deprecated tendency to cloudiness and frequent rain ; which renders our weather usually unpleasant to persons coming from a

---

＊ I have published on a broad sheet, and in anticipation of that part of my subject, a compendious account, illustrated by a diagram, of the principal phenomena of Temperature in the Climate of London. It is entitled *A Companion to the Thermometer,* and may be had of the publisher of this work.

more southern, or even from a northern clime, if belonging to a continent. Habit, however, completely reconciles the Englishman to a sky, which rarely glows for a week together with the full sun, and which *drips,* more or less on half the days of the year : and he finds, in the vigour inspired by its moderate cold, and in his mental energy, which is kept alive by its incessant changes, an indemnity for any deprivation of the listless animal enjoyment, in which the African and the Asiatic grow weary. Vicinity to the Sun's direct rays is the source of *their sameness,* and singular as it may appear, to those little conversant in such subjects, it is demonstrable, from abundant evidence, (enough of which is even contained in this volume), that we owe most of *our vicissitude,* even in temperature, to the *Moon.* It appears that our attendant planet, principally, if not solely, by the effect of gravity, continually disturbs the density of the atmosphere, producing, in the temperate latitudes of the globe, a *variety of currents,* the different qualities of which, in respect of temperature, moisture and electricity, are developed in the region over which they successively pass. Hence great *variety of weather* ;—this however, on the great scale of the year, is regulated by the more or less predominant influence of the Sun's rays, according to his place in declination : which secure to us the enjoyment of our *four seasons* in succession, these fluctuations notwithstanding.

Referring to the work at large for their varieties, let us review the seasons in their *mean* or standard state—and then the *months* in succession, in their meteorological properties and relations. It may however be repeated here, in order to complete the general standard, that the *mean height of the Barometer,* for the period from 1807 to 1816 inclusive to which my own results are referable, is 29.823 inches: that of the Royal Society exhibiting for the same period a mean of 29.849 inches.

That the mean annual Rain, at the surface, for the like period is 24.83 inches, and that both a longer average of my own, and the average of the register of the Royal Society, for the ten years previous to the above period, *when corrected for the elevation of the guage*, give a result very near to 25 inches.

I have stated that the character of the years, in other respects, follows in great degree the mean temperature. To apply this to the Rain, it will be found that from 1810 to 1816 inclusive, the *warm* years were uniformly *dry*, or below the average in rain, and the *cold* ones uniformly *wet*, or above the average. It is also re-markable that, after an extreme wet year in 1797, we had four successive years very near the average in rain, and then an extreme dry year: and that the same series appears to be now in repetition from 1816: that very wet year having been already succeeded by three average years, and the fourth, 1820, presenting for the half of it elapsed, the same character: so that there is every reason to presume that 1821 will prove an extreme dry year. Thus the Rain appears to have a cycle of increase and decrease, as well as the Tem-perature, though it may not be limited to the same extent of years as the former.

The *mean of De Luc's Hygrometer* for the climate is 66 degrees: and the character of its prevailing winds *Westerly*.

Our WINTER begins, by the temperature, the 7th of the Twelfth month, *December*, and continues 89 days, in Leap-years, 90 days.

The mean temperature of the *season* in the country,

is 37.76°* During it, the *medium* temperature of the
twenty-four hours descends from about 40° to 34½ degrees, and returns again to the former point.†

The mean height of the Barometer is 29.802 inches,
being .021 in. above that of Autumn. The range of
the column is greatest in this season; and in the
course of twenty winters it visits nearly the two ex-
tremities of the scale of three inches. The mean winter range is however 2.25 inches.

The predominating winds at the beginning of winter are the S—W: in the middle, these give place to
Northerly winds—after which they prevail again, to
the close: they are at this season often boisterous at
night.

The mean Evaporation, taken in situations which
give more than the natural quantity from the surface
of the earth (being 30.467 in. on the year) is 3.587
inches. This is a third *less* than the proportion indicated by the mean temperature; shewing the *dampness* of the air at this season.

De Luc's Hygrometer averages about 78 degrees.

The average Rain is 5.868 inches. The rain is
greatest at the commencement, and it diminishes in
rapid proportion to the end. In this, there appears a
salutary provision of Divine Intelligence: for had it
increased, or even continued as heavy as in the autumnal months, the water instead of answering the
purpose of irrigation, for which it is evidently de-
signed, would have descended from the saturated surface of the higher ground in perpetual floods, and

---

* As deduced from the averages in Table G, for the period from
1807 to 1816.

† According to the Table of averages on 20 years, p. 142—143, in
which the City observations form one half of the period. It descends
to 30.70°, on 10 years in the Country: p. 140.

wasted for the season. the fruitful plains and vallies. See on this subject the Notes under Table 28.

Notwithstanding the sensible indications of moist-ure, which in the intervals of our short frosts attend this season, the actual quantity of vapour in the at-mosphere is now, probably, at its lowest proportion. Or rather, it is so at the commencement of the sea-son ; after which it gradually increases with the tem-perature and evaporation.

In consequence of this low state of the vapour, and the generally weak electricity, in mild weather, the Clouds exhibit little variety, and are easily, and there-fore frequently, resolved into rain. The *Cirrostratus* and *Cumulostratus*, with abundance of *scud*, or the scattered rudiments of the Cumulus, chiefly appear : the whole sky hangs low, and the region below it, to the earth, is more or less *misty*. Yet we are not now wholly exempt from thunderstorms ; which occur, ap-parently in consequence of the sudden and plentiful decomposition of vapour, brought in by strong south-erly winds.

Hail is, however, of rare occurrence in our winter, if we except a sprinkling of small opake *grains*, which in the fore part of the night indicate the approach of a low temperature, and are found on the frozen ground, and on the ponds, in the morning.

The Snow crystallizes, with us, when slowly and scantily produced, in forms not so various perhaps as those of higher latitudes, yet sufficiently beautiful to be worthy, at all times, of examination : the star of six rays, carrying more or less of secondary branches at an angle of 60°, is the most common. In this re-spect also the *rime*, which collects on our trees and shrubs, when it just freezes with a moist air, presents considerable variety, and is occasionally magnificent.

The hoar-frost, which whitens our fields usually at the approach of rain, and is not confined to this season, is of two kinds. Tue most common is *spicular*, like the rime, and collected in this form from the air, though I have some doubt whether the particles are usually frozen until the moment of their attachment to the support : the other is *granular*, and consists of the drops of dew, beautifully solidified by the cold, as they rest on the herbage

Our great frosts are preceded by continued thick mists, from the condensation of the vapour, which continues for some time to be emitted by the rivers and other waters ; as well as by the moist soil, until frozen to some depth. I have gone into some detail on the phenomena of our hard and stormy winters, in the Notes under Tables 89. 90, 101, 102, 114 and 115, to which the reader is referred. The simple difference of 4° or 5°, in the medium temperature, suffices sometimes to effect the change from a damp misty state of the air, to comparative dryness and serenity—or the contrary. Our winters, therefore, present every variety of weather which can be expected within the limits of the temperature—from the calm frosty night, with its short day of chearful sunshine, to the gloomy, or thickly clouded sky, when the Southwest wind *surges* among the leafless trees through the nights ; or the more dreaded Northeast prevails through the twenty-four hours, driving the snow before it.

From the uncertain occurrence of really dangerous weather in our winters, it is probable that the people make less of the needful provision of clothing, use less foresight in their movements, and in effect, suffer more in proportion from the cold, than the inhabitants of higher latitudes.

2 p

Spring commences the 6th of the Third month, March: its duration is 93 days, during which the *medium* temperature is elevated, in round numbers, from 40 to 58 degrees. The mean of the *season* is 48.94°— the Sun effecting by his approach an advance of 11.18° upon the mean temperature of the winter. This increase is retarded in the fore part of the Spring, by the winds from North to East, then prevalent; and which form two-thirds of the complement of the season; but proportionately accelerated afterwards by the Southerly winds, with which it terminates. A strong *Evaporation* in the first instance, followed by *showers* (often with thunder and hail) in the latter, characterize this period. The temperature commonly rises, not by a steady increase from day to day, but by sudden starts, from the breaking in of sunshine upon previous cold cloudy weather. At such times, the vapour appears to be now and then thrown up, in too great plenty, into the cold region above; where being suddenly decomposed, the temperature falls back for a while, amidst wind, showers and hail, attended in some instances with frost at night.

I have given, under Table 32, a detailed account of one of these hail-storms, the ravages of which I myself witnessed. Our own island, however, suffers but little from them, compared with the fine fields of some provinces of France; which from time immemorial have been subject to their destructive visits. Human ingenuity, always exercised in one way or other in an uncertain strife with the elements, has here however resorted to a bold and singular expedient, and the French actually blow up the nascent storm with gunpowder! An account of this process, as practised in the high lands of the district of the *Maconnais*, is given under Table 6: and the same page presents an instance, in an accident at Silkstone, *Yorkshire,* (where several persons were drowned by a torrent proceeding from rain in the district above them) of what may be suffered in a neighbourhood, from the want of skilful observation of the gathering of thunderstorms, and the probable course of the waters which they may discharge, in a mountainous country.*

The heat and vapour, notwithstanding these interruptions, accumulate on the whole; and the atmosphere now receives an addition of .030 in. upon the mean of the winter—the Barometer averaging 29.832 inches. But the extreme elevations and depressions of the column go off, in great measure, during the season; and by the end of Spring the range is contracted to about an inch and a half. Mean range of the season 1.81 in.

The Evaporation, taken as before, amounts to 8.856 inches; being about a sixth part *more* than the proportion indicated by the mean Temperature.

Consistently with this proof of *dryness,* the average of De Luc's Hygrometer is 61 degrees.

The average Rain is 4.813 inches. It increases at a small rate through the season: but being greatly exceeded by the evaporation, the soil uniformly gets dried; and the light springs, which issued during the winter from the superficial strata, disappear, or become insignificant.

The lower atmosphere becomes very transparent in the fore part of the season: but the brilliancy of the

---

* In the same page, in an account of a thunderstorm which struck a flour-mill, the large chain by which the sacks were drawn up, is stated to have been " melted to a rod of iron." The fact so uncouthly stated probably was, that the tremendous charge thus conducted was just sufficient to reduce the links to a white heat; when the weight of the chain drew their sides together.

returning Sun is apt to be eclipsed, during pretty long intervals, by a close veil of *Cumulostratus* clouds; which cover the whole sky with their drapery, connected at certain points by a kind of central stem, or basis of the structure, hanging low in the sky. At other times, under the same course of Easterly or Northerly winds, there appear very regular ranks of a meagre *Cumulus,* coming on from the horizon, and passing away to the opposite quarter, with little or no change of form or magnitude, and unattended in great measure with any other modification. But in the latter part of the season we have more variety of Clouds. The *Cirrus,* which is connected with variable breezes throughout the year, now assumes more of tint and consistency, and is peculiarly fine before thunderstorms: and majestic *Nimbi* traverse the sky in succession, affording slight showers of large opake hail or snow; the prodigious electricity attending which seems to prove, that these singular clouds really act as *conductors,* fitted by communicating a portion of the repulsive fluid, to prepare the way for the descent of subsequent showers, without the necessity of explosions. See an account of the electrical phenomena of one of these, under Table 6; by which it appears that the centre of a shower is *positively* charged, while the circumference is *negative :*—a fact which at once affords a clue for explaining many of the most sudden, and apparently capricious changes, discoverable by the insulated rod, when showers are flying about in distinct bodies; the separate charges of which must independent of their own composition, produce many phenomena by affecting each other.

It is remarkable, that a *snow-storm,* in the middle of this season, not unfrequently proves the forerunner of the first hot weather, which is developed in ten days, or at most two weeks after it; consistently with this fact, some of the swallow tribe, of which different species come from the South, to avail themselves of our temperate summers for breeding, (if not also to shun the tropical rains) make their first appearance in the midst of such weather. This seems to prove that their approach is not gradual, but rather a rapid flight to our shores, by the help of a superiour Southerly current: and some observations on the phenomena consequent on their disappearance, induces me to suspect, that they avail themselves of similar aid, from a high Northerly current, to return.

A wet Spring seems not at all ungenial in our climate, provided it be followed by a warm and dry summer, as was remarkably the case in 1818: but in general, dry weather, however cold, in the early part of this season, appears to be the wish of our farmers, who have no objection to showers after they have got their seed into the ground. " Humida solstitia atque hyemes optata serenas Agricolæ," says the Latin poet; whose rules in some particulars indicate a climate not so remote from our own, as is that of Mantua at this day. But should the farmer have too much rain for the business in which he is now occupied, it may be some source of consolation to him to be satisfied (as he may be in general) that the circumstance indicates a dry time for the ensuing harvest. I have shewn, in page 203, in what years during my own observation, a portion of the rain usually belonging to the Autumnal equinox, was thus anticipated by the Vernal.

SUMMER begins the 7th of the Sixth month, *June,* and lasts 93 days. The mean temperature of the season is 60.66°, or 11.72° above that of Spring. The *medium* of the twenty-four hours rises during the season from 58° to 65°; and returns again by the close, to the former level.

The mean height of the Barometer for Summer is 29.877 inches; or .045 in. above the Vernal mean. The atmosphere now acquires, under the more vertical rays of the Sun in Full North declination, the greatest quantity of heat and vapour which it at any time contains; and it accordingly weighs most by the Barometer. The range of this instrument still diminishes to the middle of the season, when it does not exceed an inch: it then gradually increases again to the end: the mean range 1.08 in.

I have shewn that the great fluctuations in the density or gravity of the atmosphere, in our climate, are principally due to our participation, by turns, of the polar and tropical atmospheres, between which we are situate. But our position in Summer, when by the inclination of our pole towards the Sun, we are presented in a more direct manner to the rays, approximates the habits of our climate to those of the equatorial regions; and we thus become more uniform, both in temperature and density, than at any other season;—though still greatly more variable, in both respects, than the countries in that part of the globe. In proportion as the Sun rises higher, and continues longer above the horizon, the Moon, to whose influence I have attributed the variable winds of our climate, becomes depressed, as to our latitudes: her influence, consequently, is diminished, and that of the Sun, to which we have seen ascribed a more uniform action on the winds, is established in its place. Such appear to be the reasons, why the Barometer varies so much less in Summer than in winter: but its movements in ascent or descent in this season, are not therefore the less indicative of those changes in the density of the air, on which the weather, in some considerable degree, depends.

An important part of the Agricultural business of our district, *the making of hay*, is chiefly conducted within the limits of this season. I have no doubt, that this branch of rural œconomy has derived very considerable aid from the use of the Barometer—and in fact, that much less of valuable fodder is spoiled by wet, now, than in the days of our forefathers. But there is yet room for improvement in the knowledge of our farmers, on the subject of the atmosphere. It must be a subject of great satisfaction and confidence to the husbandman, to know at the beginning of a Summer, by the certain evidence of Meteorological results on record, that the season, in the ordinary course of things, may be expected to be a dry and warm one : or to find, in a certain period of it, that the average quantity of rain to be expected for the month has already fallen. On the other hand, when there is reason, from the same source of information, to expect much rain, the man who has courage to begin his operations under an unfavorable sky, but with good ground to conclude, from the state of his instruments and his collateral knowledge, that a fair interval is approaching, may often be profiting by his observations; while his cautious neighbour, who waited for the weather to " settle," may find that he has let the opportunity go by. This superiority, however, is attainable by a very moderate share of application to the subject; and by the keeping of a plain diary of the Barometer, and Rain-guage, with the Hygrometer and the vane under his daily notice.

In this respect, the rule of the distribution of Rain according to the Moon's declination, (see p. 251, &c) may be kept in mind with some prospect of advantage—the time most favorable to *hay-making* being clearly that in which the Moon is far South of the Equator. We are not yet prepared to anticipate the return of such an effect from the Full Moon, as that which we now know to have attended it in 1807. Future registers, and the examination of those extant, will however enable us to do this : and in the mean time it is well to remember, that any judgment respecting the weather is likely to be safe, in proportion only as it is founded on fixed principles, carefully applied by observation.

The predominating winds of our Summer are clearly the W—N, or those which range from the West to North, the latter point not included.

The Mean Evaporation is 11·580 inches, being above a fourth part more than the proportion indicated by the Temperature. De Luc's Hygrometer averages 52 degrees through the season.

The mean Rain is 6·682 inches. I have treated at some length, under the head Rain, of the proportions of Rain in the different seasons ; and shown the reason why, if we divide the year into two moieties by the Solstices, we have very unequal proportions of rain with nearly equal mean temperatures ; if by the Equinoxes, then very unequal temperature, for the two halves, with nearly equal proportions of rain. I have likewise, in that part of the work, endeavoured to shew the connexion of our rain with the prevailing winds ; and the different quarters from which we may more immediately derive the vapour, which forms rain in the different seasons.

Referring to the diagram, page 193, and the several divisions immediately following it, I may here shortly state, that our Summer rains, which are much the most plentiful in the middle of the season, or during the Seventh Month, appear to be the result of a less powerful and constant operation of the same causes which produce the Tropical rains. Hence our wettest summers are those, in which, by the concurring effects of the Sun's declination and the currents, we partake the most of the Tropical atmosphere: and we obtain a dry Summer only by approximating, in consequence of an opposite course of winds, or an atmosphere generally calm or breezy, to the circumstances of the high Northern latitudes. In the one case, we seem to be placed in the great general stream of subsiding air from the South; in the other, in air returning from the Northward, after having deposited its excess of water. A North-west current is therefore our fair weather wind; which will bring us moderate weather and sunshine, so long as it is not interfered with by Southerly currents, which I have shewn to arrive, when they bring us rain and thunder, for the most part with an Easterly direction; consequently in a way the most likely to mix with, and be decomposed by, the prevailing Westerly current.

When there exists a tendency to this process, our Summer clouds, in consequence of the greater quantity and more elevated situation of the vapour, exhibit a magnificence approaching to that of the Tropical sky. The *Cirrus*, which is usually the first to make its appearance after serene hot weather, now spreads its tufts to a greater extent, and assumes a more dense appearance than in spring; and the *Cumulus*, ever beautiful and of favorable aspect when insulated in the midst of sunshine, now tends constantly to inosculate above, or become grouped laterally, with the *Cirrocumulus* and *Cirrostratus* which occupy the middle region. From the mixture of these, and the inter persion of a quantity of *anomalous haze*, in patches

2 Q

or extensive beds, there results a sky, more readily
remembered than described, which is very easily and
suddenly resolved into thunder showers. The locality
of these is often determined by the place of a rapidly
growing Cumulus, which becoming a centre of union
for the surrounding looser portions, gradually extends
itself above and around, till it has put on the form of
Cumulostratus, the last stage before the explosion,
which decides the precipitation of the water in heavy
rain. This once begun, the *Nimbus*, with a confused
moving and spreading sheet, increasing the obscurity
on all sides, renders further observation from below
very imperfect. At every interval, however, of some
hours duration, with the same winds, the same state
of the sky returns again. A tendency to *rain* in such
a sky, is perhaps as decidedly indicated, by the grouping of the *Cirrus* and haze in certain parts, in the
form of the crown of a *Nimbus*, as by any other symp-
tom ; while the *Cirrocumulus*, which is the proper
natural index of a rising temperature, is favorable to
dryness ; except as it forms a part of the preparatory
machinery of thunder storms. In the latter case it is
usually arranged on a kind of arched base, mixed
with the Cirrus and Cirrostratus, and the whole with
the haze above mentioned. The immediate tendency
to an electrical explosion is always indicated, to those
who have the view of the lower part of the cloud, by
a surprisingly quick motion of the loose ragged portions of *scud* around it; which seem in haste to obey
the powerful attraction of the mass, and take their
places in the general arrangement, on which probably
the effect depends. A thunder storm in profile on
the horizon, in the dusk of the evening, is one of the
most sublime spectacles in nature. Such is the immense depth and extent, and the picturesque forms

and complex arrangement of these natural batteries, before the explosions :—and when these have commenced, it is easy, for a while, to discover the very cloud from which each proceeds, the whole substance of it becoming, at the moment, *incandescent* with electric light. In proportion as the charge is drawn off, the high-wrought forms of the clouds disappear, the crowns of Nimbi are spread out above, and the whole passes into the more familiar appearance of a distant bank of showers; which in effect it now constitutes.

Autumn begins the 8th of the Ninth month, *September*, and occupies 90 days. The mean Temperature is 49.37°—or 11.29° below the Summer: the medium of the day declines in this season from 58° to 40°

The mean height of the Barometer is 29.781 inches; being .096 in. below the mean of Summer. The range increases rapidly during this season; the mean extent of it is 1.49 inches.

The prevailing Winds are the class S–W, throughout the season.

The Evaporation is 6.444 inches, or a sixth part *less* than the proportion indicated by the Temperature. The mean of De Luc's hygrometer is 72 degrees.

The average Rain is 7.441 inches : the proportion of rain increases, from the beginning to near the end of the season : this is the true rainy season with us, and the earth, which had become dry to a considerable depth during the Spring and Summer, now receives again the moisture required for springs, and for the more deeply rooted vegetables, in the following year.

These changes in the state of the atmosphere in

Autumn are all referable to one and the same cause—
the return of the Sun to the South. The heat declining daily, the store of vapour in the atmosphere undergoes a continued decomposition, the loss of weight
arising from which is not made up, as in Summer, by
an equal production of new vapour. Hence a declining Barometer, with extensive heavy rains, chiefly in
the latter part of the season. The whole increase,
derived on the average of the Barometer in Spring
and Summer, is thus disposed of, and the atmosphere
returns to the minimum of its weight. From the
more saturated state of the air, the Evaporation falls
short of the Temperature; and the Hygrometer, at
the same mean temperature, exhibits an average 11
degrees more moist than that of Spring.

The fore part of this season is, nevertheless, if we
regard only the sky. the most delightful part of the
year, in our climate. When the decomposition of
vapour, from the decline of the heat, is as yet but in
commencement—or while the electricity remaining in
the air, continues to give buoyancy to the suspended
particles, a delicious calm often prevails for many
days in succession, amidst a perfect sunshine, mel-
lowed by the vaporous air, and diffusing a rich golden
tint, as the day declines, upon the landscape.

At this period, chiefly the *Stratus* cloud, the lowest
and most singular of the modifications, comes forth in
the evenings, to occupy the low plains and valleys,
and shroud the Earth in a veil of mist, until revisited
by the Sun. So perfectly does this inundation of suspended aqueous particles imitate real water, when
viewed in the distance at break of day, that I have
known the country people themselves deceived by its
unexpected appearance.

A phenomenon attends this state of the air, too remarkable to be passed over in silence, though it belongs, in strictness, to another branch of Natural history. An immense swarm of small spiders take advantage of the moisture, to carry on their operations ; in which they are so industrious, that the whole country is soon covered with the fruit of their labours, in the form of a fine net-work ; the presence of which I have at different times noticed in my Journal, under the term *Gossamer*. They appear exceedingly active in pursuit of the small insects, which the cold of the night now brings down ; and commence this *fishery* about the time that the swallows give it up, and quit our shores. Their manner of loco-motion is curious : half volant, half aeronaut, the little creature darts from the papillæ on his rump, a number of fine threads which float in the air. Mounted thus in the breeze, he glides off with a quick motion of the legs ; which seem to serve the purpose of wings, for moving in any particular direction. As these spiders rise to a considerable height, in very fine weather, their tangled webs may be seen descending from the air in quick succession, like small flakes of cotton.

On threads of stronger texture, produced by some of these autumnal spiders, and which I have found often extended from tree to tree, for some yards in length, the most minute dew-drops collect in close arrangement, and on the first touch of the support, run together and fall down ; giving thus a practical illustration of the manner of the formation of rain in the atmosphere. And both on these, and on webs placed in an oblique or vertical direction, on the shrubs and herbage, and formed with the symmetry usually displayed by this insect, these drops are occasionally found frozen ; and a string of little ice beads may be taken up from the web. From the texture thus covered with dew, the solar rays at times reflect innumerable little *irides,* which will not be overlooked by those who know how to appreciate the smaller beauties of an autumnal landscape.

Nor should the heavy *dews* of this season pass unmentioned; which I have sometimes found so abundant, as to be capable of daily measurement in the rain-guage. See the Notes under Tables 24 and 37

In the drops of dew, when of considerable size, under the clear morning Sun, the Meteorologist may find a good instance of the reflection and refraction which produce the Rainbow. He has only to place himself with his back to the Sun, and singling out a particular drop which appears brilliant with any colour, he may, by changing his position, so as to vary the angle, and keeping his eye on the drop, draw out the different prismatic tints in succession.

The latter part of this season, and beginning of Winter, are more peculiarly subject to gales of wind from the Southwest. While our Northeast breezes are plainly the result of sunshine, and blow almost exclusively by day *these* appear to prevail chiefly by night: the one forming part of an ascending, the other of a subsiding set of currents. That our Westerly gales come from above, is manifest from the manner in which the clouds indicate, before-hand, the increase and decrease of velocity which they afterwards manifest below. And there seems no way of accounting for their occasional excessive force, but by attributing it to the Westerly *momentum,* which the air acquires in a higher latitude, by revolving in a larger circle about the Earth's axis, and which it may bring with it when suddenly translated Northward. It is even worthy of consideration, whether the sudden depressions of the Barometer, of a few hours duration only, which accompany these gales, and exhibit their minimum about the time of the greatest force of the wind are not to be attributed *to an actual loss of gravity, by the centrifugal force in the air, for the time.*

However much we may be exposed, at seasons, to these boisterous visitants, the tremendous concussion of the elements, properly denominated a *hurricane* is almost unknown in our climate. Yet we may not conclude it absolutely exempt from real hurricanes. The force of the wind is occasionally such for a short time, as to give it the characteristic qualities of this phenomenon—in uprooting and breaking timber trees, damaging solid buildings, and rolling up or removing the heavy sheets of lead with which they are covered. See Tables 28, 39, 77, 101.

To produce, however, one of the most memorable instances on record, I shall make some extracts from an old publication, having the following title, The Storm, or a collection of the most remarkable casualties and distresses which happened in the late dreadful tempest both by sea and land. London, 1704. 12 mo. p. 272. The motto, The Lord hath his way in the whirlwind, and in the storm, and the clouds are the dust of his feet. Nah. i. 3.

The date of this tempest, as to its extreme violence, is the night of the 26–27 November, O.S. 1703, being about the time of New Moon. It appears to have been preceded by a very wet season for about six months.

" It had blown exceeding hard (says the anonymous compiler, who seems to have been a man of respectable rank, and careful to have authorities for his facts) for about fourteen days past, that we thought it terrible weather: several stacks of chimnies were blown down, and several ships lost, and the tiles in many places blown off the houses; and the nearer it came to the fatal 26th of November, the tempestuousness of the weather encreased.

" On the Wednesday morning before, being the 24th of November, it was fair weather and blew hard; but not so as to give any apprehensions, till about four o'clock in the afternoon the wind increased, and with squalls of rain and terrible gusts blew very furiously. The wind continued with unusual violence all the next day and night; and had not the *Great Storm* followed so soon, this had passed for a great wind.

" On Friday morning it continued to blow exceeding hard, but not so as that it gave any apprehensions of danger without doors. Towards night it increased; and about 10 o'clock our Barometers informed us that the night would be very tempestuous—*the mercury sunk lower than ever I had observed it on any occasion*. It did not blow so hard till 12 at night, but that most families went to bed: but about 1, or at least by 2 o'clock, 'tis supposed, few people that were capable of any sense of danger, were so hardy as to lie in bed. And the fury of the tempest increased to such a degree, that as the Editor of this account, being in London and conversing with the people the next day, understood, most people expected the fall of their houses. And yet, in this general apprehension, nobody durst quit their tottering habitations; for whatever the danger was within doors, 'twas worse without. The bricks, tiles, and stones, from the tops of the houses, flew with such force, and so thick in the streets, that no one thought fit to venture out, though their houses were near demolished within.

" It is the received opinion of abundance of people, that they felt, during the impetuous fury of the wind, several movements of the earth; and we have several letters which affirm it: but as an earthquake must have been so general, that every body must have discerned it; and as the people were in their houses when they imagined they felt it, the shaking of which might impose upon their judgment, I shall not venture to affirm it was so. Others thought they heard it thunder. 'Tis confess'd the wind by its unusual violence made such a noise in the air, as had a resemblance to thunder;—the roaring had a voice as much louder than usual, as the fury of the wind was greater than was ever known: the noise had also something in it more formidable; it sounded aloft, and roared not very much unlike remote thunder. And yet, though I cannot remember to have heard it thunder, or heard of any that did, in or near London, in the countries, the air was seen *full of meteors and vaporous fires*; and in some places both thunderings and universal flashes of lightning, to the great terror of the inhabitants.

" From two of the clock the storm continued, and increased till five in the morning; and from five to half an hour after six, it blew with the greatest violence. The fury of it was so exceeding great for that particular hour and half, that if it had not abated as it did, nothing could have stood its violence much longer. In this last part of the time, the greatest part of the damage was done. Several ships, that rode it out till now, gave up all—for no anchor could hold. Even the ships in the Thames were all blown away from their moorings, and from Execution-dock to Limehouse hole, there were but four ships that rode it out: the rest were all driven down into the *Bight,* as the sailors call it, from Bell wharf to Limehouse, where they were huddled together and drove on shore, heads and sterns, one upon another, in such a manner as any one would have thought it had been impossible.

" The points from which the wind blew are variously reported from various hands. 'Tis certain, it blew all the day before at SW, and I thought it continued so till about two o'clock; when as near as I could judge by the impressions it made on the house—for we durst not look out—it veer'd to the SSW, then to the W; and about six o'clock to W by N— and still the more Northward it shifted the harder it blew, till it shifted again Southerly about seven o'clock; and as it did so it gradually abated.

Though the storm abated with the rising of the Sun, it still blew exceeding hard; so hard that no boats durst stir out upon the river, but upon extraordinary occasions; and about three o'clock, the next day, it increased again, and we were in a fresh consternation.—At four it blew an extreme storm with sudden gusts as violent as at any time of the night; but

as it came with a great black cloud and some thunder, it
brought a hasty shower of rain which allayed the storm, so
that in a quarter of an hour it went off, and only continued
blowing as before.

" This sort of weather held all Sabbath-day and Monday,
till on Tuesday afternoon it increased again, and all night
blew with such fury that many families were afraid to go to
bed.—At this rate it held blowing till Wednesday about one
o'clock in the afternoon, which was that day sevennight on
which it began—so that it might be called one continued
storm from Wednesday noon to Wednesday noon. In all
which time there was not one interval, in which a sailor
would not have acknowledged it blew a storm; and in that
time two such terrible nights as I have described."

Such a tempest could not be supposed to be limited
to this island—accordingly it appears to have spread
over a great part of the North of Europe, though no
where with equal impetuosity as with us. As to the
effects, they were generally these : Over most part of
South Britain and Wales, the tallest and stoutest timber trees were uprooted, or snapt off in the middle.
It was computed that there were twenty-five parks in
the several counties, which lost a thousand trees apiece
—the New Forest, Hants, above four thousand. Whole
sheets of lead were blown away from the roofs of
strong buildings : seven steeples, above four hundred
windmills, and eight hundred dwelling houses, blown
down ; and barns, out-houses, and ricks in proportion
besides a great destruction of orchards. About one
hundred and twenty persons lost their lives on land,
among whom were the Bishop of Bath and Wells and
his lady, who unhappily lodged in a ruinous castle :
also the engineer who had erected the then lighthouse
at the Eddystone ; who was blown into the sea along
with the structure, which he had promised himself
would bid defiance to the elements.

At sea there were few ships to sink—the previous terrible weather having brought them into port in very unusual numbers—but in the harbours and roadsteads of England, so many vessels ran foul of each other and sunk, or foundered at anchor, or were driven on the sands, or to sea where they were never heard of, that it is computed eight thousand seamen at least perished on the occasion. A vessel laden with tin, being left in the small port of Helford near Falmouth, with only a man and two boys on board, drove from her four anchors at midnight, and going to sea, made such speed before the wind, almost without a sail, that at eight in the morning, by the presence of mind of one of the boys, she was put into a narrow creek in the Isle of Wight, and the crew and cargo saved.

This run may give us some conception of the velocity of the wind: for if we consider that the course of the vessel, even by the winds, could not have been direct, but in a large curve outwards from the coast, the rate at which she went exceeded thirty miles an hour on the average: and that of the wind must have been three or four times greater.

The estuary of the Severn, lying more particularly in the course of this storm, the parts bordering on that river suffered much by the breaking in of the sea. The country for a great extent was inundated, the vessels driven upon the pasture land, and many thousands of sheep and cattle drowned.

To conclude this description, the spray of the sea was on this occasion carried far inland in such quantities, as to form little concretions or *knobs* of salt on the hedges; and at twenty-five miles from the sea, in Kent, made the pasture so salt, that the cattle for some time would not eat it. The total damage was

considered, by the Editor of the work I have been
quoting, to exceed that of the great fire of London.

First Month. January.

The Sun in the middle of this month continues
about 8 h. 20 m. above the horizon. The *Temperature*
rises in the day, on an average of twenty years, to
40·28° ; and falls in the night, in the open country,
to 31·36°—the difference 8·92°, representing the
mean effect of the Sun's rays for the month, may be
termed the *Solar variation* of the Temperature.

The Mean temperature of the month, if the observations in the city be included, is 36·34° But this
mean has a range, in 10 years, of about 10 25°, which
may be termed the *Lunar variation* of the Temperature. It holds equally in the decade, beginning with
1797, observed in London, and in that beginning
with 1807, in the country. In the former decade, the
month was coldest in 1802, and warmest in 1804: in
the latter, it was warmest in 1812, and coldest in
1814.* I have likewise shewn in the curves facing
page 144, that there was a tendency in the *daily*
variation of temperature through this month, to proceed, in these respective periods of years, in opposite
directions. The prevalence of different classes of
winds in the different periods, is the most obvious
cause of these periodical variations of the mean
temperature.

The *Barometer* in this month rises, on an average

---

* As the Temperature of the month in other years approaches near
to either extreme, the reader will do well to consult Table A, from
whence these results are taken, as he proceeds.

of ten years, to 30·40 in. and falls to 28·97 in. the *mean range* is therefore 1 43 in.; but the extreme range in 10 years is 2·38 in. The mean height for the month is about 29·79 inches.

The prevailing *winds* are the class from West to North. The Northerly predominate, by a fourth of their amount, over the Southerly winds.

The average Evaporation (on a total of 30·50 inches for the year) is 0·832 in., and the mean of De Luc's Hygrometer 80.

The mean Rain, at the surface of the earth, is 1·959 in.; and the number of days on which snow or rain falls, in this month, averages 14,4.

A majority of the *nights* in this month have constantly the temperature at or below the freezing point.

Second Month.   February.

Length of day in the middle of the month about 9 h. 55 m.

Mean of greatest heat by day 44·63°, of greatest cold by night 33·70°: difference, or *Solar variation* 10·93°

Mean temperature of the month, the city temperature included, 39·60°: difference in the mean, or *Lunar variation,* from 1797 to 1806, in London, 7·45°; from 1807 to 1816, in the country, 11·75°. The month was coldest in 1800 and 1814, and warmest in 1806 and 1809.

The Barometer ranges, on a mean, from 30·42 to 29·07 in.: difference 1·35 in.: but the full range in 10 years extends to 2·01 inches. Mean height for the month 29·874 in.

The prevailing winds are the class from South to West.

The proportionate evaporation is 1·647 in.: and the mean of De Luc's Hygrometer 75 degrees.

The average Rain at the surface is 1·482 in.: and the average number of days on which any falls, 15,8. This is the month in which, on the whole, rain or snow falls the oftenest. The frosty nights vary from three to twenty-two, and the average of these on ten years is *eleven*.

Third Month.   March.

The middle day has the Sun for about 11 h. 50 m.

The mean heat rises by day to 48·08°, and falls by night to 35·31°; the *Solar variation* is therefore 12·77°.

The Mean temperature of the month is 42·01°, the London observations included. The *Lunar variation*, in the first decade, was 6·74°; in the second 11·08°   The month was coldest in 1799 and 1807, and warmest in 1801 and 1815.

The mean range of the Barometer in this month is from 30·40 to 29·10, or 1·30 in.; the full range for 10 years being 1·80 in.   Mean height for the month 29·87 inches.

The prevailing winds are decidedly the class from North to East: and these sensibly impede the advance of the Temperature.

The proportionate evaporation is 2·234 in.: and the mean of De Luc's hygrometer 67 degrees.

The average Rain at the surface is 1·299 in.: and it rains on a mean, in this month, only on 12,7 days.

There are, on an average, *twelve* frosty nights in this month; the proportion varies from five to twenty-three.

Fourth Month. April.

The Sun is above the horizon, in the middle of the month, about 13 h. 57 m. The Temperature rises by day to 55·37°, and falls by night to 39·42° : the *Solar variation* is consequently 15·95°.

The Mean temperature of the month, for London and the country, is 47·61°. It varied in the first decade 7·54°, in the second 8·64° : the *Lunar variation* is therefore probably pretty uniform. The month was warmest in 1798 and 1811, and coldest in 1799 and 1808.

The Barometer ranges, on an average, from 30·23 to 29·15 inches : the mean range is therefore 1·08, but the full range in ten years is 1·62 inches. Mean height for the month 29·814 inches.

The prevailing winds are still from North to East : yet the advance of the Temperature is now somewhat quicker.

The proportionate evaporation is 2·726 in : being little more than in last month : the mean of De Luc's Hygrometer is however 60.

The mean amount of Rain at the surface is 1·692 in.: and it falls on an average on 14 days of this month.

During ten years, this month never passed quite without a frosty night or morning, and it has *six* of these on an average.

Fifth Month. May.

The length of the middle day is about 15 h. 35 m. The heat rises, on a mean, in the day to 64·06°, and falls in the night to 46·54°. *Solar variation* 17·52°.

The Mean temperature of the month, with the city included, is 55·40° : the *Lunar variation* in the first decade was 7·44° ; in the second 10·54°. The month was warmest in 1804 and 1811, and coldest in 1802 and 1816.

The Barometer rises, on a mean, to 30·25, and falls to 29·34, mean range 0·91 in : but the full range in ten years is 1·52 inch.   Mean height for the month 29·812 inches.

The prevailing winds are the class from South to West: by means of which, in aid of the sun, we get an advance in this month, upon the last, of near eight degrees, and acquire some heat against the coming in of summer.

The proportionate evaporation is now 3·896 in.: and the mean of De Luc's Hygrometer 57 degrees.

The average rain at the surface is 1·822 in.: and rain falls on 15,8 days of this month.

In five seasons out of ten, the nocturnal Temperature, or that a little before sun-rise, touches in this month once or twice upon the freezing point.

Sixth Month.   June.

The length of day, in the middle of this month, extends to about 16 h. 32 m.   The mean heat rises in the day to 68·36°, and falls in the night only to 49·75°: the *Solar variation* is then 18·61°, which is the largest for the year; consistently with the Sun's greatest altitude on the 21st of this month.

The Mean temperature of the month, with London included, is 59·36°   It varies, in the ten years from 1797 to 1806, 6·44°, and in the following decade 5·80°   The *Lunar variation* is therefore pretty uniform, and on the whole at its minimum for the year. The month was warmest in 1798 and 1811; and coldest in 1797 and 1812.

The Barometer rises, on a mean, to 30·28, and falls to 29·45; the difference 0·83 in.: but the full variation in this month, for ten years, is 1·25 inch: the mean height for the month about 29·90 inches.

The prevailing winds are the class from West to North.

The proportionate evaporation is 3·507 inches, being less than that of last month, the advance of the temperature notwithstanding. The mean of De Luc's Hygrometer 52 degrees.

The mean amount of rain is 1·920 in.: the average number of days on which any falls is 11·8 only, being the lowest of any month in the year.

I meet with no instance, in the course of ten years, of a frosty night or morning in this month.

Seventh Month. July.

Length of the middle day about 16 h. 5 m. Mean highest temperature by day 71·50°; mean lowest by night 53·84°. *Solar variation* 17·66°.

Mean temperature of the month, for London and the country, 62·97°. The *Lunar variation* is nearly uniform, being 7·14° for the decade of observations in London, and 8·40° for that in the country. The month was hottest in 1803 and 1808, and coldest in 1802 and 1812.

The Barometer rises, on the average, no higher than 30·18, and sinks only to 29·49 inches: the mean range is therefore 0·69 in.: and the full range being only 0·99 in. it has in this month the smallest variation in the year. Mean height about 29·88 inches.

Notwithstanding this state of the Barometer, the class W—N continues to include the winds most prevalent in this month: and these, with their antagonists from the Southward, occasionally blow with considerable force at this season. See the Notes under Table 121.

2 s

The proportionate evaporation is 4·111 inches; and the mean of De Luc's Hygrometer again 52 degrees.

The average rain of the month is 2·637 inches: and rain falls, on a mean, on 16 days of this month.

Eighth Month. August.

On the middle day of this month we have the Sun for about 14 h. 32 m. The heat rises by day, on a mean, to 71·23°, and sinks by night only to 53·94°: we have now the warmest nights in the year, and but little abatement of the temperature by day. The *Solar variation* therefore keeps up to 17·29°; which is a degree and one-third more than in the corresponding month in spring, the fourth. But we are to recollect that, at the present season of the year, the action of the Sun's rays is considerably assisted by the warm earth, which radiates heat into the air: while in spring, it absorbs every day a proportion of the heat which the Sun produces.

The mean heat of the month, with London included, is 62·90°. The *Lunar variation* is very uniform, being 7·07° in the first decade, and 7·44° in the second. The month was hottest in 1802 and 1807, and coolest in 1799 and 1812.

The Barometer rises on a mean in this month to 30·19, and sinks to 29·43 in; the mean range is consequently ·76 in: the full range for 10 years is 1·02 in. Mean height for the month 29·854 in.

The winds from West to North prevail most in this month also.

The proportion of Evaporation is 3·962 inches, and De Luc's hygrometer continues to average 52 degrees.

The average Rain of the month is 2·125 inches; and it falls, on a mean, in this month, on 16·3 days. The rain in this, and the two preceding summer months, presents a very uniform average, (when corrected for the different elevation of the guages,) on the two decades of years.

Ninth Month. September.

The middle day is about 12 h. 39 m. from sun-rise to sun-set. The heat, on a mean, rises to 65·66°, and falls to 48·67°; making a *Solar variation* of 16·99°. This is but a fraction of a degree less than that of last month, and 4·22° more than in the Third month. From the different place of the Equinox in each, the Third is astronomically colder than the Ninth month; but we have besides, and in a greater degree, a source of inequality in their temperature, in the absorption and radiation of heat by the earth, already treated of.

The Mean temperature of this month is, for London and the country together, 57·70°. The *Lunar variation* is nearly uniform, being 6·61° for the first decade, and 5·98° for the second. The month was coldest in 1803 and 1807, and warmest in 1804 and 1810.

The mean of ten maxima of the Barometer for this month is 30·23 in: of ten minima 29·33 in: giving a mean range of ·90 in: but the full range on ten years is 1·54 in.: mean for the month 29·883 in.—the greater oscillations beginning now to come on again, in proportion as the temperature declines, and the currents get more interchange of direction, Northward and Southward. The prevailing winds of this month are, on the whole, the class S—W

The mean proportionate evaporation is 3·068 inches, and the mean of De Luc's Hygrometer 64 degrees.

The average Rain for the month is 1·921 in: and the number of days on which any falls, only 12,3: so that, on the ten years from 1807, it is the month which stands next, in point of dryness to the sixth; the Third being, however, nearly equal to it. It is proper to remark, that on the decade from 1807 to 1816, the average rain for this month is about half an inch *less* than on the former decade; while for the Third month, it is about as much *more*. The translation of a portion of rain from the autumnal to the vernal equinox obtained, therefore, in a greater degree in the latter, than in the former period.

In this month we have occasionally a frosty night or two. In 1816, it froze pretty sharply on the 2d: in 1815, on the 6th and 7th: and in 1807, by placing a Thermometer near the ground, I detected a Temperature of 26°, on the 13th of the month.

Tenth Month.   October.

The middle day of this month has the Sun for about 10 h. 37 m. The mean of greatest heat by day is 57·06°, of greatest cold by night 43·51°: the *Solar variation* 13·55°.

The mean Temperature, for the city and country, is 50·79°: the *Lunar variation* of which is smaller, in the decade for the city, than in any other instance in these observations, being only 4·51°: but in the country decade it amounts to 9·18°. I am not prepared to suggest any peculiar cause for this very small variation in one decade, while a mean one obtains in the other; but we shall see a more remarkable instance of this in the Twelfth month. The Tenth month was coldest in 1797 and 1814, and warmest in 1804 and 1811.

The Barometer in this month rises, on a mean, to 30·21, and sinks to 29·05 in.: mean range 1·16, full range 1·82 in: mean height for the month 29·736 in.

The Winds from South to West predominate.

The proportionate Evaporation is 2·208 in.: and De Luc's Hygrometer averages now 71 degrees.

The average Rain is 2·522 inches: it rains, on a mean, on 16·2 days of the month. The month became somewhat wetter in the decade beginning with 1807, than it had been in the preceding ten years.

The Tenth month is less subject to frost, than we might expect from the advanced period of the year. It has not usually above *four* nights, and sometimes none at all, with the Thermometer below the freezing point: the warmth of the ground is one obvious cause.

Eleventh Month. November.

The middle day of the month extends only to 8 h. 49 m. The average temperature, however, rises to 47·22°, and sinks only to 36 49°, making a *Solar variation* of 10·73°; while in the First month, with a mean day of 8 h. 20 m. it was but 8·92°. It may be necessary, both in this month and the last, to admit as an additional cause of the comparative warmth, the heat given out by the great quantity of aqueous vapour now condensed into rain: while in Spring, the temperature may be proportionately kept down by the effect of evaporation, in which process much of the atmospheric warmth occasionally disappears.

The mean Temperature, for the whole district, is 42·40°. It varies 7·52° in the decade beginning with 1797, and, 8·14° in that beginning with 1807; the *Lunar variation* is therefore nearly uniform. The month was coldest in 1798 and 1816, and warmest in 1806 and 1811.

The Barometer, which constantly enlarged its movements in receding from the Summer season, now exhibits the greatest *depressions*. It rises, on a mean, to 30·36, and sinks to 28 90 inches, making the range in this way 1·46 in. : but the full range in ten years is 2·12 inches. Mean height for the month 29·725 inches.

Consistently with this state of the Barometer, we have now the Southwest winds oftenest—but with a large mixture at intervals of Northerly ; which average considerably above their mean. Perhaps the greatest interchange of these currents now takes place in our atmosphere.

The proportionate Evaporation is now reduced to 1·168 in. ; and the mean of De Luc's Hygrometer advanced to 80 degrees.

The mean Rain of this month is 2·998 inches : it is consequently the wettest month in the year : and it is observable, that it was somewhat drier in the decade beginning with 1807, than in the preceding ten years. Rain falls on precisely half the number of days in the month.

In this, and the preceding month, but most in the present, the depressions of temperature occur, which bring on the cold of Winter in our climate. It will be seen that we lose about 8·degrees on the mean. A gloomy windy sky is accordingly the prevailing characteristic of the season ; but this is not constant ; and we have at intervals also in this month, very fine days, with clear nights and hoar-frosts.

With regard to frosts, this month has eleven or twelve nights on an average, on which the Thermometer is at or below 32° ; and the following *gradation* appears in the number of such nights in the month, from 1811 to 1816 inclusive, viz. 5,.9, 12, 14, 17, 20 ; shewing the progressively increasing tendency to a low temperature, as this series of years proceeded.

Twelfth Month. December.

Our day is reduced, in the middle of this month, to about 7 h. 46 m. The average Temperature rises to 42·66°, and sinks to 33·90° The *Solar variation* is therefore only 8·76°, being the smallest for the year; consistently with the Sun's place in declination.

The mean Temperature, with London included, is 38·71°. It varied, in the decade from 1797 to 1806, to the extent of 14·45°; and in that from 1807 to 1816, only 5·45° The *Lunar variation* is therefore now the most extensive, and the least uniform. The mean of this month in 1799, in London, was 34·30°: the month had become nearly as cold in the preceding year, having then sunk in its mean about $7\frac{1}{2}$ degrees. In 1806 it was, by the London observations, 48·75°, and by my own at Plaistow 46·27°. It had increased in the preceding year; and a similar gradation from cold to warmth in this month took place in the years 1801–2–3, when it was carried in the last, to 42·78°. In the latter decade, on the contrary, the gradation twice proceeds, through four years, towards the lower extreme.

By returning to Tables 1, 2, the reader will see the *immediate* cause of the hibernal warmth, in the great prevalence of Southerly winds, more especially the SW. By these, the temperature of the fore part of the winter was kept up nearly to the pitch of spring: and the Notes present some instances of the effects which this had on vegetation. A decided flow of air from the Southward, without almost the intervention of a frosty night, must be expected to produce a very different mean, from a season in which most of the nights are frosty. And the difference from this cause be expected to be greatest, when we have the least of the Sun's influence: but the question, what *it is that determines this Southerly current to our district,* in one year or series of years more than another, remains to be solved.

The Barometer rises, on a mean, to 30·40, and sinks to 28·95 inches, giving a mean range of 1,45 in. : the extreme range being 2·37 inches, nearly as large as in the First month, which is the greatest for the year. The mean height of the Barometer is 29·745 inches.

The Westerly winds, on the whole, preponderate. The proportionate Evaporation is 1·12 in. : and De Luc's Hygrometer continues to exhibit its winter mean of 80 degrees.

The average Rain is 2·427 inches. Rain or snow falls on 1,77 days of the month ; which is therefore, nearly, the most subject to what is sometimes provincially termed " falling weather."

On an average of ten years, about half the nights in this month appear to be frosty.

TABLE OF LUNATIONS.

FOR THE WHOLE SPACE OCCUPIED BY THE REGISTER.

Compiled from WHITE'S EPHEMERIS.

N. B.—The minutes are always in *addition* to the hour; the mark (a) denotes that the *time expressed* is between midnight and noon, and (p) that it is between noon and midnight.

Year	New Moon day h. m.			First Quarter day h. m.			Full Moon day h. m			Last Quarter d y h. m.		
1806	Nov. 10	11	41 a.	18	7	20 a.	26	2	1 a.	Dec. 2	10	44 p.
	Dec. 10	2	24 a.	18	4	31 a.	25	3	p			
1807										Jan. 1	6	57 a.
	Jan. 8	7	36 p.	16	12	26 a.	24	2	31 a	30	5	18 p.
	Feb. 7	2	15 p.	15	5	35 p.	22	12	47 p.	Mar. 1	6	21 a.
	March 9	8	52 a.	17	7	5 a.	23	10	9 p.	30	9	45 p.
	April 8	2	7 a.	15	4	38 p.	22	7	15 a.	Apr 29	2	39 p.
	May 7	5	6 p.	14	10	59 p.	21	4	42 p	May 29	8	3 a.
	June 6	5	24 a.	13	3	30 a.	20	3	10 a.	June 28	1	15 a.
	July 5	3	15 p.	12	7	53 a.	19	3	11 p.	July 27	5	46 p.
	Aug. 3	11	27 p.	10	1	52 p.	18	5	10 a.	Aug. 26	9	4 a.
	Sept. 2	7	5 a.	8	10	46 p.	16	9	6 p.	Sept. 24	10	34 p.
	Oct. 1	3	13 p.	8	11	21 a.	16	2	30 p.	Oct. 24	9	56 a.
	31	12	42 a.	Nov. 7	3	45 a.	15	8	14 a	Nov. 22	7	16 p.
	Nov. 29	11	58 a.	Dec. 6	11	21 p.	15	12	55 a.	Dec. 22	3	17 a.
	Dec. 29	1	9 a.									
1808				Jan. 5	8	54 p.	13	3	31 p.	20	11	6 a.
	Jan. 27	4	9 p.	Feb. 4	6	31 p.	12	3	53 a.	18	7	47 p.
	Feb. 26	8	43 a.	March 5	1	55 p.	12	2	21 p.	19	5	53 a.
	Mar. 27	2	11 a.	April 4	5	28 a.	10	11	26 p.	17	5	36 p.
	Apr. 25	7	28 p.	May 3	4	42 p.	10	7	39 a.	17	7	2 a.
	May 25	11	19 a.	June 1	12	23 a.	8	3	34 p.	15	10	8 p.
	June 24	12	56 a.	July 1	5	45 a.	7	12	3 a.	15	2	53 p.
	July 23	12	18 p.	30	10	21 a.	Aug. 6	10	5 a.	14	8	41 a.
	Aug. 21	10	10 p.	Aug. 28	3	40 p.	Sept. 4	10	41 p.	13	2	30 a.
	Sept. 20	7	27 a.	Sept. 26	10	56 p.	Oct. 4	2	18 p.	12	7	6 p.
	Oct. 19	4	54 p.	Oct. 26	9	10 a.	Nov. 3	8	27 a.	11	9	41 a.
	Nov. 18	2	55 a.	Nov. 24	11	3 p.	Dec. 3	3	35 a.	10	9	52 p.
	Dec. 17	1	36 p.	Dec. 24	4	43 p.						
1809							Jan. 1	9	53 p.	9	7	51 a.
	Jan. 16	1	9 a.	23	1	23 p.	31	2	8 p.	Feb. 7	4	13 p.
	Feb. 14	1	59 p.	22	11	2 a.	Mar. 2	3	57 a.	Mar. 8	11	43 p.
	Mar. 16	4	19 a.	24	7	18 a.	31	3	23 p.	April 7	7	9 a.
	Apr. 14	7	57 p.	22	12	27 a.	Apr. 30	12	41 a.	May 6	3	26 p.
	May 14	12	4 p.	22	1	55 p.	May 29	8	18 a.	June 5	1	30 a.
	June 13	3	42 a.	20	11	58 p.	June 27	3	7 p.	July 4	2	2 p.
	July 12	6	13 p.	20	7	25 a.	July 26	10	14 p.	Aug. 3	5	21 a.
	Aug. 11	7	33 a.	18	1	20 p.	Aug. 25	7	3 a.	Sept. 1	11	4 p.
	Sept. 9	7	58 p.	16	6	51 p.	Sept. 23	6	38 p.	Oct. 1	6	10 p.
	Oct. 9	7	42 a.	16	1	12 a.	Oct. 23	9	25 a	31	1	32 p.
	Nov. 7	6	49 p.	14	9	40 a.	Nov. 22	2	57 a.	Nov. 30	7	18 p.
	Dec. 7	5	21 a.	13	9	17 p.	Dec. 21	10	p.	Dec. 29	10	47 p.

TABLE OF LUNATIONS.

Year	New Moon			First Quarter			Full Moon			Last Quarter		
	day	h.	m.	day	h.	m.	day	h.	m.	day	h.	m.
1810 Jan.	5	3	37 p.	12	12	32 p.	20	5	6 p.	28	11	14 a.
Feb.	4	2	8 a.	11	6	53 a.	19	10	58 a.	26	8	37 a.
Mar.	5	1	23 p.	13	2	47 a.	21	2	31 a.	28	3	39 a.
Apr.	4	1	37 a.	11	10	32 p.	19	3	8 p.	26	9	28 a.
May	3	2	46 p.	11	4	41 p.	19	1	10 a.	25	3	25 p.
June	2	4	38 a.	10	8	20 a.	17	8	18 a.	23	10	47 p.
July	1	7	6 p.	9	9	9 p.	16	2	50 p.	23	8	40 a.
	31	10	10 a.	Aug. 8	7	20 a.	14	9	46 p.	21	9	43 p.
Aug.	30	1	35 a.	Sept. 6	3	24 p.	13	6	17 a.	20	2	5 p.
Sept.	28	4	46 p.	Oct. 5	10	14 p.	12	5	6 p.	20	9	17 a.
Oct.	28	6	58 a.	Nov. 4	4	57 a.	11	6	29 a.	19	6	8 a.
Nov.	26	7	44 p.	Dec. 3	12	43 p.	10	10	20 p.	19	2	47 a.
Dec.	26	7	9 a.									
1811				Jan. 1	10	30 p.	9	4	16 p.	17	9	11 a.
Jan.	24	5	45 p.	31	10	56 a.	Feb. 8	11	27 a.	16	12	3 p.
Feb.	23	4	3 a.	Mar. 2	1	56 a.	Mar. 10	6	18 a.	17	11	3 p.
Mar.	24	2	12 p.	31	6	56 p.	Apr. 8	11	3 p.	16	6	48 a.
Apr.	22	12	19 a.	Apr. 30	1	3 p.	May 8	12	39 p.	15	12	26 p.
May	22	10	42 a.	May 30	7	12 a.	June 6	11	7 p.	13	5	16 p.
June	20	10	2 p.	June 28	12	18 a.	July 6	7	26 a.	12	10	43 p.
July	20	11	4 a.	July 28	3	34 p.	Aug. 4	2	53 p.	11	6	7 a.
Aug.	19	2	12 a.	Aug 27	4	42 a.	Sept. 2	10	35 p.	9	4	39 p.
Sept.	17	6	57 p.	Sept. 25	3	48 p.	Oct. 2	7	16 a.	9	7	1 a.
Oct.	17	12	9 p.	Oct. 25	1	16 a.	31	5	19 p.	Nov. 8	1	16 a.
Nov.	16	4	28 a.	Nov. 23	9	36 a.	Nov. 30	7	9 a.	Dec. 7	10	24 p.
Dec.	15	7	11 p.	Dec 22	5	30 p.	Dec. 29	7	11 p.			
1812										Jan. 6	8	19 p.
Jan.	14	8	18 a.	21	1	49 a.	28	11	38 a.	Feb. 5	4	40 p.
Feb.	12	8	p.	19	11	27 a.	27	5	51 a.	Mar. 6	9	49 a.
Mar.	13	6	22 a.	19	11	p.	27	12	16 a.	April 4	11	5 p.
Apr.	11	3	27 p.	18	12	41 p.	26	5	10 p.	May 4	8	37 a.
May	10	11	39 p.	18	4	16 a.	26	7	34 a.	June 2	3	10 p.
June	9	7	50 a.	16	9	7 p.	24	7	33 p.	July 1	7	53 p.
July	8	5	13 p.	16	2	25 p.	24	5	45 a.	30	12	18 a.
Aug.	7	4	55 a.	15	7	28 a.	22	2	59 p	Aug. 29	6	1 a.
Sept.	5	7	22 p.	13	11	39 p.	20	11	51 p.	Sept. 27	2	32 p.
Oct.	5	12	10 p.	13	2	25 p.	20	8	51 a	Oct. 27	2	53 a.
Nov.	4	6	14 a.	12	3	15 a.	18	6	30 p.	Nov. 25	7	20 p.
Dec.	3	12	20 a.	11	1	51 p.	18	5	23 a.	Dec. 25	3	7 p.
1813												
Jan.	2	5	21 p.	9	10	27 p.	16	6	4 p.	24	12	34 p.
Feb.	1	8	36 a.	8	6	2 a.	15	8	43 a.	22	9	44 a.
Mar.	2	9	30 p.	9	1	43 p.	17	12	48 a.	25	4	46 a.
Apr.	1	7	55 a.	7	10	28 p.	15	5	20 p	23	8	25 p.
	30	4	14 p.	May 7	8	54 a.	15	9	26 a.	23	8	8 a
May	29	11	21 p.	June 5	9	17 p.	14	12	32 a.	21	4	16 p.
June	28	6	26 a.	July 5	11	40 a.	13	2	24 p.	20	9	57 p
July	27	2	43 p.	Aug. 4	4	1 a.	12	2	57 a.	19	2	43 a
Aug.	26	1	8 a.	Sept. 2	10	p	10	2	13 p.	17	8	8 a

TABLE OF LUNATIONS.

Year	New Moon day h. m.		First Quarter. day h. m.		Full Moon day h. m.		Last Quarter day h. m.	
1813	Sept. 24 2 11 p.	Oct. 2 4 46 p.			10 12 31 a.		16 3 34 p.	
	Oct. 24 5 56 a.	Nov. 1 10 58 a.			8 10 23 a.		15 2 1 a.	
	Nov. 22 11 58 p.	Dec. 1 3 3 a.			7 8 25 p.		14 3 53 p.	
	Dec. 22 7 15 p.	30 4 10 p.						
1814					Jan. 6 7 8 a.		13 9 3 a.	
	Jan. 21 2 13 p.	29 2 22 a.			Feb. 4 6 46 p.		12 4 45 a.	
	Feb. 20 7 11 a.	27 10 26 a.			Mar. 6 7 15 a.		14 1 30 a.	
	Mar. 21 9 6 p.	28 5 21 p.			April 4 8 29 p.		12 9 23 p.	
	Apr. 20 7 55 a.	26 12 6 a.			May 4 10 29 a.		12 2 41 p.	
	May 19 4 23 p.	26 7 31 a.			June 3 1 15 a.		11 4 27 a.	
	June 17 11 32 p.	24 4 33 p.			July 2 4 34 p.		10 2 54 p.	
	July 17 6 26 a.	24 4 3 a.			Aug. 1 7 51 a.		8 10 54 p.	
	Aug. 15 2 5 p.	22 6 46 p.			30 10 26 p.		Sept. 7 5 34 a.	
	Sept. 13 11 18 p.	21 12 41 p.			Sept. 29 11 53 a.		Oct. 6 11 58 a.	
	Oct. 13 10 51 a.	21 8 49 a.			Oct. 28 12 16 a.		Nov. 4 7 4 p.	
	Nov. 12 1 15 a.	20 5 16 a.			Nov. 27 11 52 a.		Dec. 4 3 45 a.	
	Dec. 11 6 36 p.	19 12 5 a.			Dec. 26 11 p.			
1815					Jan. 2 2 52 p.			
	Jan. 10 1 57 p.	18 4 2 p.			25 9 47 a.		Feb. 1 5 2 a.	
	Feb. 9 9 31 a.	17 4 44 a.			23 8 16 p.		Mar. 2 10 8 p.	
	Mar. 11 3 21 a.	18 2 19 p.			25 6 37 a.		April 1 5 7 p.	
	Apr. 9 6 20 p.	16 9 21 p.			23 5 17 p.		May 1 12 17 p.	
	May 9 6 20 a.	16 2 47 a.			23 4 57 a.		31 6 4 a.	
	June 7 3 53 p	14 7 53 a.			21 6 p.		June 29 9 41 p.	
	July 6 11 47 p.	13 2 12 p.			21 8 33 a.		July 29 11 2 a.	
	Aug. 5 6 57 a.	11 11 13 p.			19 12 11 a.		Aug. 27 10 22 p.	
	Sept. 3 2 21 p.	10 12 noon			18 4 14 p.		Sept. 26 7 57 a.	
	Oct. 2 10 55 p.	10 4 44 a.			18 8 3 a		Oct. 25 4 8 p.	
	Nov. 1 9 34 a.	9 12 34 a.			16 11 8 p.		Nov. 23 11 32 p.	
	30 10 51 p.	Dec. 8 9 50 p.			16 12 58 p.		Dec. 23 7 9 a.	
	Dec. 30 2 51 p							
1816		Jan. 7 6 41 p.			14 1 18 a.		21 4 13 p.	
	Jan. 29 8 50 a.	Feb. 6 1 29 p.			13 12 9 p.		20 3 42 a.	
	Feb. 28 3 31 a.	Mar. 7 4 55 a.			13 9 47 p.		20 5 41 p.	
	Mar. 28 9 27 p.	April 5 4 22 p.			12 6 43 a.		19 9 38 a.	
	Apr. 27 1 31 p.	May 4 12 8 a.			11 3 40 p.		19 2 35 a.	
	May 27 3 7 a.	June 3 5 18 a.			10 1 19 a.		17 7 48 p.	
	June 25 2 7 p.	July 3 9 28 a.			9 12 21 p.		17 12 46 p.	
	July 24 11 9 p.	31 2 25 p.			Aug. 8 1 18 a.		16 4 58 a.	
	Aug. 23 7 6 a.	Aug. 29 9 43 p.			Sept. 6 4 22 p.		14 7 47 p.	
	Sept. 21 3 3 p.	Sept. 28 8 25 a			Oct. 6 9 19 a.		14 8 35 a.	
	Oct. 20 11 56 p.	Oct. 27 10 58 p.			Nov. 5 3 18 a.		12 7 8 p.	
	Nov. 19 10 23 a.	Nov. 26 5 6 p.			Dec. 4 8 51 p.		12 3 52 a.	
	Dec. 18 10 37 p.	Dec. 26 1 52 p						
1817					Jan. 3 12 44 p.		10 11 42 a.	
	Jan. 17 12 38 p.	25 11 43 a.			Feb. 2 2 15 a.		8 7 46 p.	
	Feb. 16 4 19 a.	24 8 27 a.			Mar. 3 1 35 p.		10 4 53 a.	
	Mar. 17 9 11 p.	26 2 2 a.			Apr. 1 11 9 p.		8 3 28 p.	
	Apr. 16 2 28 p.	24 3 23 p.			May 1 7 33 a.		8 3 39 a.	

TABLE OF LUNATIONS.

Year	New Moon day h. m.	First Quarter day h. m.	Full Moon day h. m.	Last Quarter day h. m.
1817	May 16 7 a.	24 12 42 a.	30 3 21 p.	June 6 5 37 p.
	June 14 9 45 p.	22 7 4 a.	June 28 11 18 p.	July 6 9 25 a.
	July 14 10 17 a.	21 11 56 a.	July 28 8 22 a.	Aug. 5 2 51 a.
	Aug. 12 9 p.	19 4 50 p.	Aug. 26 7 36 p.	Sept. 3 9 2 p.
	Sept. 11 6 43 a.	17 11 3 p	Sept. 25 9 47 a.	Oct. 3 2 42 p.
	Oct. 10 4 15 p.	17 7 44 a	Oct. 25 2 55 a.	Nov. 2 6 43 a.
	Nov. 9 2 8 a.	15 7 44 p.	Nov. 23 9 56 p.	Dec. 1 8 21 p.
	Dec. 8 12 33 p.	15 11 29 a	Dec. 23 4 59 p.	31 7 33 a.
1818	Jan. 6 11 36 p.	14 6 44 a.	22 10 26 a.	29 4 42 p.
	Feb. 5 11 38 a.	13 4 2 a.	21 1 29 a.	27 12 27 a.
	March 7 12 59 a.	15 1 8 a.	22 2 1 p.	29 7 37 a.
	April 5 3 44 p.	13 7 51 p.	20 12 13 a.	27 3 4 p.
	May 5 7 26 a.	13 11 8 a	20 8 29 a.	26 11 49 p.
	June 3 11 13 p.	11 10 51 p.	18 3 28 p.	25 10 46 a.
	July 3 2 18 p.	11 7 37 a.	17 10 14 p.	25 12 33 a.
	Aug. 2 4 22 a.	9 2 23 p.	16 6 5 a.	23 5 12 p.
	31 5 28 p.	Sept. 7 8 9 p	14 4 14 p.	22 11 56 a.
	Sept. 30 5 48 a.	Oct. 7 2 5 a.	14 5 29 a.	22 7 30 a.
	Oct. 29 5 28 p.	Nov. 5 9 24 a.	12 9 49 p.	21 2 29 a.
	Nov. 28 4 27 a.	Dec. 4 7 19 p	12 4 19 p.	20 7 32 p.
	Dec. 27 2 52 p.			
1819		Jan. 3 8 40 a	11 11 36 a.	19 9 43 a.
	Jan. 26 1 10 a.	Feb. 2 1 24 a	10 6 15 a.	17 8 39 p
	Feb. 24 11 53 a.	Mar. 3 8 31 p.	11 11 2 p.	19 4 41 a
	Mar. 25 11 24 p.	April 2 4 19 p.	10 1 5 p.	17 10 47 a
	Apr. 24 11 48 a.	May 2 11 13 a.	9 12 6 a.	16 4 17 p
	May 24 1 2 a.	June 1 4 7 a.	8 8 30 a.	14 10 34 p
	June 22 3 1 p.	30 6 28 p.		

NOTE.—A scheme of the weather has been for some years in circulation under the title of *Herschel's Table* (disclaimed however by the celebrated Astronomer of that name) in which the changes are made to depend, in great measure, on the hour of day or night, at which the Moon enters upon her several phases—*noon* being the point most likely to be followed by rain, and *midnight* most favourable to fair weather: the meridian for which it is calculated not expressed. This scheme appears to me to be empirical, and its merits consequently very dubious—but the reader who may possess and incline to examine it, as to London, may do this by comparing the times of the Lunations, as here stated, with the *Rain column* and Notes in my Tables; referring to the scheme as he proceeds.

TERMINOLOGY OF CLOUDS, &c.

In the Introduction, page xxxii, I promised an explanation of the terms used for Clouds. I shall therefore extract, from the Essay already in print on that subject, the *definitions* of the several kinds of Clouds; referring to that piece, for a more particular account of their nature. To this I shall add a few definitions of other phenomena, mentioned in these volumes, with such remarks as may assist a person, already somewhat conversant in Meteorology, to distinguish them with accuracy.

Modifications of Clouds.

Clouds are susceptible of various modifications.

By this term is intended the structure or manner of aggregation, in which the influence of certain constant laws is sufficiently evident, amidst the infinite less diversities resulting from occasional causes.

Hence the principal modifications are as distinguishable from each other, as a tree from a hill, or the latter from a lake; although clouds, in the same modification, compared with each other, have often only the common resemblances which exist among trees, hills, and lakes, taken generally.

There are three simple and distinct modifications, which are thus named and defined.

1. Cirrus. *Def.* Nubes cirriformis tenuissima, quæ undique crescat.

The Cirrus. A cloud resembling a lock of hair, or a feather. Parallel flexuous, or diverging fibres, unlimited in the direction of their increase.

2. Cumulus. *Def.* Nubes densa cumulata, sursum crescens.

The Cumulus. A cloud which increases from above in dense, convex, or conical heaps.

3. Stratus. *Def.* Nubes strata, aquæ modo expansa, deorsum crescens.

The Stratus. An extended, continuous, level sheet of cloud, increasing from beneath.

There are two modifications, which appear to be of an intermediate nature ; these are :

4. Cirrocumulus. *Def.* Nubeculæ subrotundæ connexæ vel ordinatè positæ.

The Cirrocumulus. A connected system of small roundish clouds, placed in close order, or contact.

5. Cirrostratus. *Def.* Nubes extenuata, sub-concava vel undulata. Nubeculæ hujusmodi appositæ.

The Cirrostratus. A horizontal or slightly inclined *sheet,* attenuated at its circumference, concave downward, or undulated. Groups or patches having these characters.

Lastly, there are two modifications, which exhibit a compound structure, viz.

6. Cumulostratus. *Def.* Nubes densa, quæ basi cumuli structuram patentem cirro-strati vel cirrocumuli superdat.

The Cumulostratus. A cloud in which the structure of the Cumulus is mixed with that of the Cirrostratus, or Cirrocumulus. The Cumulus flattened at top, and overhanging its base.

7. Nimbus. *Def.* Nubes densa, supra patens et *cirriformis,* infra in pluviam abiens.

The Nimbus. A dense cloud, spreading out into a crown of Cirrus, and passing beneath into a shower.

After the experience of eighteen years, I do not find a necessity either to make additions to this little system, or to retrench any of its parts. Some subordinate distinctions may, indeed, at a future period, be found useful; but until the classification, as it is, be generally adopted, its simplicity must form its most powerful recommendation; next to (what I trust it has been found to possess) a strict conformity to nature. For like reasons, I am averse from giving on every occasion, *plates* of the modifications; having found that the copying of a particular *form* of each tends too much to limit the reader's views, in looking for it in the sky; while the exhibition of many varieties, would only serve to perplex him at the outset. It is in the *structure,* carefully considered with due reference to situation, that he will find the basis of a correct judgment; and he will do well to wait, at first, for several successive appearances of each modification, to which he will thus at length find the definition apply in all its parts.

Remarks on the Modifications.

1. The *Cirrus* comprehends every thing that is manifestly *fibrous* in structure, among the clouds. In the Nimbus it exists *in composition,* forming the crown or summit of the cloud: the connexion with falling rain in this case, being generally obvious, from an obscurity beneath the cloud, as seen above the horizon.

The Cirrus occupies commonly the highest place in the sky; but is sometimes observed mixed with, and even occasionally below, other modifications. Its density and progressive motion are also, usually, the least of any: though the propagation of its fibres by increase is occasionally pretty rapid. It may go on

increasing, both in extent and density, to a completely
overcast sky; in which case, it is followed by other
modifications below, and ultimately by rain.

This cloud may be regarded as a collector of the
Electricity, as well as of the scattered particles of
water, from the higher atmosphere. In this respect
its nature is very well illustrated by the arborescent
figures, which Electricians form, by projecting a
coloured powder on cakes of wax, previously touched
with the knob of a charged phial.

2. The *Cumulus*, on the contrary, is an aggregate,
which seems to retain an electric charge just sufficient
for its own buoyancy, during its increase. The
several clouds of this kind must, likewise, exercise a
degree of repulsive action on each other, since when
the modification is pure, they keep as it were at measured distances in the sky. In this state also, each
cloud would always be a spheroid, or roundish aggregate, were it not for the evaporation prevailing in
the lower atmosphere at the time, which cuts off all
the Cumuli visible at once to a certain height, presenting their bases in the same horizontal plane.

This is a fair-weather cloud, produced by day, and
carried, at whatever elevation, by the current which
flows next the earth. If not disposed of in one of the
compounds hereafter mentioned, it evaporates in the
evening.

3. The *Stratus* must always be looked for *in contact with the earth or water ;* in vallies, on extensive
plains, and in the moist meadows of hilly tracts. It
should not be confounded with sheets of cloud, which
may be seen buoyant in the air, resting on hill tops,
or sweeping the plains near the ground, or creeping
along the flanks and through the defiles of mountains.
All these are referable to the Cirrostratus. The pre-

sent modification does not *travel*, but is propagated, by a pretty rapid increase through a certain space and elevation, in a calm atmosphere. In the morning, it is either dissipated, by virtue of a rising temperature in the region it occupies, or driven upward with the appearance of the nascent Cumulus. It is electric, and though capable, in the course of a night, of depositing a very copious dew, does not readily wet the cloaths or the person, in passing through it.

4. The *Cirrocumulus* occupies the next place, in point of elevation and buoyancy, to the Cirrus; and is apparently the immediate result of the cessation of the conducting process, which obtains during the increase of the latter. The previous appearance of a Cirrus does not, however, appear necessary to its origin. This cloud may be considered as a system of small Cumuli, maintaining a sort of equilibrium among themselves in their electric charge, or tending slowly to inosculate laterally; in consequence, perhaps, of which tendency, the larger nubeculæ are generally seen to occupy the central part of the group. It is the natural index of a rising temperature, in the higher atmosphere; and the warmth of the region in which it floats is commonly felt, after some time, at the earth's surface. Before thunder-storms, it has a peculiar character. The nubeculæ then grow rapidly upward, upon a kind of arched base, connecting the group at bottom. Independent of its connexion with thunder, it is a fair-weather cloud; and when, as sometimes happens, it inosculates with a Cumulus beneath, retaining its rounded forms, the indication is a favorable one. It is often seen mixed in the same sky with Cirrus and Cirrostratus, and may pass to either of these modifications; the prognostic being found to vary accordingly.

5. The *Cirrostratus* is the natural index of depression of temperature in a vaporous atmosphere, and
consequently of wind and rain. A slight appearance
of it very commonly attends the production of the
evening dew. It may be conceived to originate in
the condensation of vapour in successive horizontal
strata, the product immediately subsiding through the
air below, where it still receives an increase.

It is apparently non-electric or but very weakly
charged, and wets all bodies alike. The trees on
hilly tracts drip profusely during its passage, even
while the soil remains dusty beneath; and in riding
through it, I have repeatedly found it collect on the
palpebræ, and run into the eyes. When the temperature is about the freezing point, this cloud affords
also a copious product of rime, which collects chiefly
on the windward side of the twigs and branches.

I have found it impracticable to confine this modification to the form of a perfectly continuous sheet,
although the admission of nubeculæ makes it difficult
always to separate the Cirrostratus from the Cirrus
and Cirrocumulus; more especially as these three
modifications are liable, by changes in the medium, to
pass each into the other A proper degree of attention, from time to time, to the connexion in which it
is found, and the attendant state of the atmosphere,
will nevertheless, at length, enable the observer to do
this to his satisfaction.

As a striking proof of the harmony which pervades
the works of an All-wise Creator, it may be remarked,
that while the Cirrocumulus, and the fair-weather
clouds in general, are highly beautiful, the present
modification, which presides over the wintry flood,
and the ravages of the tempest, exhibits few out of
its many forms that can be deemed in any degree

shapely; and the blots and ragged patches, in which
it sometimes abounds in a stormy season, agree perfectly with the roar of winds, and the aspect of the
denuded landscape under a frowning sky. Its ex-
tenuated edges, however, preserve it from an absolutely heavy and uncouth appearance; and in spring
and autumn, the rising or setting sun confers occasionally upon its waved streaks, in profile near the
horizon, a richness of colouring on which the eye may
dwell with pleasure.

Whether from a peculiarity of internal constitution,
or from its general thinness and continuity, this cloud
furnishes, almost exclusively, the skreens in which
are represented the circles of the *halo,* with their
occasional accompaniments, the parhelion and paraselené.

6. Of the *Cumulostratus.* This may be regarded
as the intermediate state between the Cumulus and
rain; or rather as that in which the clouds may
remain, for many hours together, prepared either to
afford showers, or to disperse, according to the course
of temperature, electricity and other attendant circumstances. It is accordingly by far the most common cloud in changeable, and even fine weather.

The Cumulus can seldom preserve its original
character long, when other vapour, besides that which
results from the superficial evaporation, is present in
excess. In such a case, acted on probably by ex-
traneous attractions, it either expands spontaneously,
as it rises, into a spreading crown, with occasional
irregular protuberances, like smoke issuing from a
furnace; or it inosculates with the superior modifications, already formed and descending upon it; or,
lastly, by lateral increase, amidst Cirrostrati floating
in the same region, or loose portions of its own kind,

which it can no longer assimilate by its electrical
attraction, its nature is changed, and a compound
ensues, the various forms of which are comprised in
the present modification.   Banks and ranges of cloud
therefore, of every description, presenting a flattish
surface, and bounded by perpendicular or overhanging
cliffs (for such they often are in real dimension) are
referable to the Cumulostratus.   Wherever, in short,
amidst the light and shade which diversify the magnificent, and often truly mountainous scenery of the
sky, that which the artist terms a *recess tint* is discoverable, it may be presumed to result from the
combinations peculiar to this modification.   The trees
and towers, ruins and glaciers, natural bridges, and
other varieties of scenery which one may fancy in the
clouds, are in general parts of the same arrangement.

A Cumulostratus, well formed and seen single in
profile, is quite as beautiful an object as the Cumulus.
Its form may be compared in general to that of a
Boletus, or to a fungus with a very thick stem and
protuberant gills: though I have occasionally seen
specimens, constructed almost as finely as a Corinthian
capital, the summit throwing, like that, a well defined
shadow upon the parts beneath.

7 The *Nimbus* is the cloud in which the electricity
of the atmosphere is most fully manifested.   In the
other modifications, it is either gradually accumu-
lating, until it may become strong enough to *super-
induce* the earth's surface, or some neighbouring collection of vapours, and thus escape directly or by
combination; or it is working its way horizontally
through extensive strata, the changes occurring in
which visibly betray its movements; but, here, it has
already opened itself a passage, and is pouring down,
(perhaps also at the same time ascending in a counter

current) in great abundance. I shall not here enlarge on the electrical phenomena of this cloud, which are noted in different parts of my observations.

This modification is evidently not *necessary* to the production of rain, any further than as an excess of electricity may require to be previously disposed of; or in other terms, the equilibrium between the earth's surface and the atmosphere to be restored. Hence it is seen so often, about the commencement and termination of rainy periods. It is scarcely possible for an accurate observer, who has once seen the Nimbus in action, afterwards to confound it with any other appearance. A central portion of dense cloud (in which by the sudden action peculiar to electrical phenomena, the large drops are formed) surmounts a column of falling rain, and is itself crowned with more or less of *cirrosity*, which in some cases extends through a great space upward and all around in straight fibres; and in others exhibits a more compact and fleecy appearance, but without losing the character of the Cirrus. These are the proper constituent parts of this modification; which however is seldom without its appendages. The Cirrostratus flanks it, in the region where the drops are formed, and Cumuli, or portions of the Cumulostratus, are seen successively to enter into this focus, whence they never emerge, but are converted into rain.

From its connexion with local showers, this cloud is distinguished, almost exclusively, by bearing, in its broad field of sable, the honours of the *Rainbow*, that most pleasing, and (were it not of so common occurrence) astonishing exhibition of nature's simple magnificence; which the believer in Divine revelation contemplates with double pleasure, when he reflects, that to it He who formed the world, was pleased, at the close of its great catastrophe, to attach the character of a perpetually recurring sign, that He would no more overwhelm it with the watery element.

Of Rain, Winds, Water-spouts, and Electrical phœnomena, connected with Clouds.

The phraseology in general use for these is at present rather confused, and requires its terms to be fixed with greater precision.

1. The term *Nimbus*, for a cloud giving rain, is clearly authorised by the best Latin writers. Then, as to the rain, &c. *Pluvia* (which seems to have been at first the adjective pluvia, used with aqua) may be appropriated to local rain from a definite mass of clouds : whilst Imber, *Imbres*, may denote *the rains*, or a more general and extensive precipitation of water: Tonitrus, tonitrua, *thunder* (as to the sound :) fulgor, *lightning :* fulmen, a *stroke of thunder*. What are called *Fire-balls*, in the accounts commonly given of thunder-storms, I believe are merely the appearance of the body of electric matter, when it moves so slowly that it may be distinctly seen in its passage, which is not commonly the case.

2. *Ecnephias*, (from *ek nephos*, Gr. *proceeding from a cloud*) seems appropriate to the sudden *gusts* which precede and accompany showers and thunder storms ; as well as to the more dangerous occasional winds of the same kind, which our seamen denominate *squalls*, and the French *coups de vent :* though in this language they seem to have a sea-term, *bouillar*, applicable to the wind and rain in connexion. Our seamen also speak of black and white, or *wet* and *dry* squalls : and it is obvious that a cloud, moving swiftly along with a column of rain, snow or hail under it, must

occasion appearances and movements in the air, differing according to the position of the observer, with respect to the shower and its course. Should the term I propose be adopted as a generic one, it may be thus defined ; Ecnephias : Ventus fortis, breve et subito ex Nimbo erumpens. Some attendant circumstances might be added, to denote the species ; thus, the phenomena described under Tab. 32 would be, Ecnephias, cum magnâ grandine tonitruis et fulgoribus : and we might have Ecnephias *pluvia,* a squall with rain, *sicca,* a dry or white squall, *nivosa, grandinosa,* and possibly other modifications.

3. *Turbo :* a whirlwind. *Tourbillon,* Fr. This term might likewise be made generic, and variously modified by additions in the manner of the last The phenomenon described under Tab. 94, and denominated by the relator a *Tornado,* would then be, Turbo, cum maximâ grandine, tonitruis et fulgoribus.

I suppose the present, as well as the last mentioned case, of a strong wind combined with the highest manifestations of electrical action, to be the result of a peculiarly rapid and abundant condensation of vapour above the clouds : in consequence of which, there is not time afforded for the regular, balanced arrangements of a thunder-cloud : which might give repeated discharges to the earth, and thus relieve the higher atmosphere. On the contrary, there is probably a sudden rushing down of the electric fluid, together with the vapour, through the imperfect conductors which may be present ; and which are assisted in their office by a further, almost instantaneous, condensation during the passage. It is obvious, if we consider the nature of electrical attraction, that in this case the course of the fluid may determine that of the air ; and make it flow during the discharge, either in a violent narrow stream horizontally, with a counter-current above or beside it; or in two whirlwinds, the one descending from the clouds upon the superinduced surface, the other ascending to restore the equilibrium. And even in the latter case, the horizontal movement of the phenomenon may be accounted for, by its speedily saturating the first spot on which it falls, which then repelling it, the cloud moves to another, and so on, in a line or track of the breadth of the cloud, and extended in length, according to the time taken by the continuous discharge.

4. *Typhon.* Def. Maximus ventorum fortium concursus. *Ouragan,* Fr. A hurricane: what is probably more strictly the application of the term *tornado*; when the wind, having blown with extraordinary violence from one quarter for a while, shifts to another, and so on, till its violence abates. See Tab. 84, and II. 29. This genus, again, may be made to comprehend different modifications, by short descriptive epithets. Hurricanes are almost always attended, I believe, with electrical manifestations, and occasionally with a great variety of them. In an account which I possess in manuscript (I know not whence extracted) of the tremendous one in the West Indies, in 1772, the writer says, " I must still mention how dreadful every thing looked in this, in itself, horrible and dark night; there being so many fiery meteors in the air, which I and others who were in the same situation were spectators of. Towards the East, the face of the heavens presented to our view a number of *fiery rods* (electrical brushes?) which were through the whole night shooting and darting in all directions; likewise *fiery balls* (bolides?) which flew up and down, here and there, and burst into a number of small pieces, and flew to and fro like torches of straw,

and came very near where we lay in the road. This was the state of the air over the town : in other parts, *another sort* of fiery balls flew through the air with great rapidity; and notwithstanding all these phenomena, common *thunder* and *lightning* was abundantly great." This account is dated, West end of Santa Cruz (Antigua) and signed, M. Smith. I have no reason to doubt its authenticity.

5. *Trombus.* The Waterspout. *Trombe,* Fr.

Were it proved that this phemomenon is the mere result of a Whirlwind, it might be denominated Turbo aquosus : but I think the accounts we have of spouts clearly prove them to be, in general, of electrical origin. I suppose the surface of the sea to be superinduced by a great mass of cloud, or condensing vapour, above; yet under such circumstances, that the necessary apparatus for regular successive discharges cannot be formed. The surface, then, rises in large papillæ, and froths up to meet the cloud, while the latter is propagated downwards in a lessening cone, to meet the water—and, the passage being thus opened, a portion of the sea water suddenly rushes up to be dispersed in the cloud, while a copious condensation of the water of the latter is carried in a stream into the sea : the rapidity of the movement, and the resistance of the air, causing one or both streams to assume a whirling motion, in order to effect their passage. The inosculation of a dense Cumulus with a sheet of Cirrostratus above it, which is a common phenomenon, often ending like the Waterspout in rain, may illustrate these more remarkable effects of a stronger charge, existing in a cloud suspended over the sea : and I should define the Trombus thus, Mutua inter aquas ad maris superficiem, et in nube proximâ, viribus electricis

motas, penetratio. If there be in nature a phenomenon, which might be fairly brought in argument, in
favour of the existence of *two fluids* in electricity, it
is, I think, this of the Waterspout. It is not always,
however, the complete electrical operation here described ; and as a thunderstorm is on different occasions a different process, with certain common appearances attending it, so the case may be in this
instance.

Of some luminous phenomena connected with Clouds.

1. *Anthelion.* Imago solis, a nube oppositâ, quasi
ab aquæ superficie, reflexa.

This is, in effect, the whole history of the Anthelion,
that, when the surface of a dense cloud presents properly for the effect, there is the same tendency to
reflect the Sun's image, as in a surface of water.
Since I first distinctly recognised the phenomenon, I
have often been able to trace it, in a greater or less
degree of perfection, on the perpendicular sides, or
in the recesses between the crown and the foot, of a
large Cumulostratus ; the cloud being opposite to the
Sun at a moderate elevation, and the sky overhead
clear. Here we may discover, at intervals, a broad
spot of light, much brighter than the rest of the cloud,
and proceeding now and then to a momentary circular image, which is presently lost again by the
increase, or change of direction, of that part of the
surface. This, I have no doubt, is the same kind of
reflexion which, under favourable circumstances, has
occasionally produced the Sun's image, amidst surrounding clouds in shade, in such brightness as to fix
the attention of even a casual observer. See p. 11.

2. *Parhelion*. Solis species falsa, diversos inter halones in nebulâ effulgens. Plerumque duo vel plures, unà cum ipso Sole emicant.

Having had few opportunities of seeing *parhelia*, I can only state, that the phenomenon appears to be seated in the points of intersection of different halos, and to derive its brightness from the union of their different reflexions in those parts. A mist near the earth, of very moderate density, surmounted by several Cirrostrati differently inclined to the horizon, may furnish the medium for a perfect exhibition of parhelion : and a frozen state of the particles composing these clouds, is perhaps accessary to the effect. It is a phenomenon which our best observers do not know when to look out for : otherwise we should probably have had a more satisfactory account of its nature and mode of production.

3. *Paraselené*. Lunæ species falsa, inter diversos halones in nebulâ visa.

I have never seen the *paraselené*, but have ventured to make its definition accord with that of parhelion ; to which the same remarks are probably applicable. See p. 19.

4. *Corona*. Area lucida, sphæram referens, quæ in nube vel nebulâ circa Lunam nocte videtur.

The Corona is so common an appearance, when thin stratified clouds are carried over by night, the Moon shining, that I have seldom thought it needful to make a note of it. A circular space full of a mild whitish light surrounds the Moon's disk ; and by the passing of the light to some degree of colour (yellow or brownish) at the outer part, its appearance may be compared to a spherical lantern, with the luminary in the midst of it. The phenomenon however varies much, and is only occasionally splendid or conspicuous.

5. *Halo.* Area lucida, inter duo vel plures circulos inclusa, plerumque colores iridis referens, quæ circa Solem aut Lunam in nube vel nebulâ videtur.

The Halo is less common than the Corona, and the Solar much less frequent than the Lunar halo. It is a broad circle of variable diameter, sometimes white, but more often exhibiting the prismatic colours; which appears in a thin cloud (or in a low diffused haze by the help of the cloud's reflexion), around the Sun's or Moon's disk. Sometimes more than one circle or series of colours appears at once, and at very different distances from the luminary in the centre.

Coronas and halos, from their connexion with the modification *Cirrostratus,* in which they chiefly appear, are found to indicate wind and rain; sometimes at the approach of winter, snow and frost; when they are high coloured; and again, late in the spring, I have repeatedly observed a large white Lunar halo to be followed merely by hot weather.

6. *Iridula.* Area circularis in qua colores iridis emicant, guttulis roris, super gramina vel aranea compestria, sparsis reflexae.

I have ventured to form this diminutive of the rainbow, in order to apply it as a generic term to those little representations of the bow, which are commonly met with in autumn, formed by the dew-drops on the grass, the gossamer, &c.

7. *Iris.*

This term may be made to comprehend at least three modifications.

1. Iris *Arcus pluvius*: the Rainbow: of which the more rare kind, exhibited in a shower by the light of the Moon, may be distinguished by the addition of the epithet *nocté* or *nocturnus.* See Tab. 26. II. 35, 68.

It is unnecessary to define this very common phenomenon. I may just remark here, that an opinion lately advanced by Dr. Watt in Thomson's Annals, that the refraction of the rays takes place, not in the raining cloud, but on the edge of another, situate between it and the Sun, appears to me to be altogether unfounded.

2. Iris *Arcus nebulæ*: the white or colourless bow, seen in a mist without rain. See Tab. 30, 75.

3. Iris *Gloria. Def.* Umbrâ spectatoris in nubem projectâ, circulos, colores iridis referentes, quasi circum caput suum pictos, in nube videt. *A glory.*

This curious phenomenon is well described, and figured, in a paper by Dr. Haygarth, in the Manchester Society's Memoirs, vol. iii. p. 463; as he observed it in the year 1780, in the vale of Clwyd. I should not, however, have thought of introducing this definition, had it not fallen to my lot to see it myself this summer.

On the 29th of the Seventh month, 1820, at Folkstone, Kent, the day was fine, with the Barometer at 30 inches, and the wind Easterly. There was a mist, of the kind which I commonly refer to the *Cirrostratus*, resting the whole forenoon on the cliffs towards Dover, and on the high land North of the town. Towards evening, the mist subsided from the cliffs, and appeared on the sea below them; a body of cloud, which appeared to be *Cumulostratus*, shewing itself also close to the horizon, on the high land abovementioned.

About half past six, p. m. walking with my family towards Sandgate, West of the town, we perceived that the mist on the sea was advancing and spreading itself Westward, and towards the shore; and a body of it came at length close under the sandy cliff, on

2 x

which we stood, at the height of about 140 feet from
the sea. This mist was of various depths: a brig
near the shore was at intervals completely hidden by
it, up to her topmasts: it exhibited a mixture of
Cumulus and *Cirrostratus.* In this state of things,
the Sun shining clear above the Western horizon,
our shadows were projected, together with that of the
cliff's edge, upon the cloud beneath, on the surface
of which, at the same time, each person could perceive, around the upper part of one of the shadows
(which being distant were small, and rather indistinct)
a luminous *corona,* surrounded by two faintly coloured
halos. The outer halo was very large, compared with
our shadows: it surrounded the whole group, and a
considerable part of the circle was cut off by the
shadow of the cliff. Consequently, when one of the
party removed to a distance, his shadow was seen to
pass the circle and appear by itself, without the *glory;*
notwithstanding which *he* continued to perceive the
whole of the phenomenon for himself, around his own
shadow; those of the rest appearing to him at a distance, and also without it. We were able to continue
these observations for about twenty minutes; until,
the Sun approaching the horizon, the shadows became
too distant to be perceived, and the circles vanished.
A thunderstorm followed these appearances, in the
night of the 30th, after which we had again fine weather. The whole phenomenon was highly curious
and interesting; and the facility with which each of
the party could either appropriate the *glory* to himself
or share it with the company present, suggested to
me some reflexions of a *moral* nature—in which,
however, I shall not anticipate the reader.

INDEX.

N. B. The references to Vol. I. are either to the *Introduction*, by the page, or to the *Tables*, (T) by the number—the nature of the subject shewing, in the latter case, whether it is to be found in the Table, or in the annexed *Notes*, &c. or in both. Those to Vol. II. are mostly by the page, but sometimes by the Table as before.

Abyssinia, account of the rains in, II. 201
Ætna, eruption of, T. 64
Anemometer, design for, by Kirwan, Int. viii
Animals, of the prognostics afforded by, T. 29
Anthelion, II. 11, 342
Aphis the cause of honey dew, T. 8
Askesian Society, II. 223
Astronomical temperature different from the real, II. 131
Atmosphere, constitution and surface of the, II. 270
————— tides in the, II. 272
Aurora borealis, T. 4, 5, 93. II. 4, 6, 70
Autumn, its duration, mean temperature, &c. II. 130, 307
Avalanches, T. 27. II. 7, 31, 27

Balloons used to ascertain currents, Int. viii. T. 37, 60, 64
Barometer, of the, Int. xi. II. 145, 153
————— its use in a storm at sea, T. 61
————— its mean height at London, II. 146, 293
————— its Yearly range and extremes, II. 147
————— its Monthly range and extremes, II. 150
————— true place of its changeable point at London, II. 153
————— why higher in Summer than in Winter, II. 283
————— why least variable in Summer, II. 302
————— wavering or unsteady, T. 28, 39, 41, 141
————— high, T. 2, 4, 15, 16, 41, 52, 53, 55, 77, 80, 113, 114, 125, 129, 151. See Tables C & F
————— low, T. 1, 6, 13, 26, 27, 28, 39, 42, 50, 51, 61, 62, 74, 87, 90, 101, 104, 105, 113, 115, 124, 126, 127, 128, 134, 138, 140, 153. See Tables C & F
————— of oil, how constructed, Int. xiv
Barometrical clock, of the, Int. xiii
Bats, T. 5, 30, 43
Beaufoy, Col. II. 290

INDEX.

Bees out at the approach of frost, T. 89, 101. II. 33
Beccaria, Int. xxxvii
Bells, electrical, of De Luc, Int. xxxi. T. 54, 55, 56
Bertholon, Int. xxxvii
Bevan, Silvanus II. 243
Birds breeding in a warm winter, T. 2
—— singing by moonlight in the heat, T. 21
—— ——— less heard after a hard winter, T. 94
—— their wings fettered by sleet, T. 28
Blue tint of the Landscape in intense cold, T. 89
Bombay, great rain at II. 31
Bruce, Journal of, at Gondar, kept by Balugani, II. 201
Burney, Dr. W. II. 68

Cat, domestic, fond of cockchaffers, T. 9
Cavallo, Int. xxx
Centrifugal force of the air may diminish its pressure, II. 311
Chain welded by a stroke of lightning, T. 6
Cirrus cloud, *passim*
——————— defined. II. 329
——————— gradation of tints in the, T. 60
——————— its appearance before thunder, T. 57
———————————————— wet weather, T. 7, 111
———————————————— a Northerly storm, T. 3
———————————————— a Southerly storm, T. 113, 114
Cirrocumulus cloud, *passim*
—————— defined, II. 330
—————— indicates heat and thunder, T. 32, 35, 48, 49, 51, 57,
    58, 59, &c.
Cirrostratus cloud, *passim*
————— defined, II. 329
————— its effect in stopping heat, T. 125
————— prognostic of rain and wind, T. 20, 32, 46, 49, 59, 61,
    62, 97, 99, 106, 108, 111, 124. II. 43
Clouds, of the terms used for, Int. xxxii. II. 229—338
———— appearance resembling combustion in the, T. 67
———— ———————— of before thunder, T. 32. II. 305
———— coloured, examples of, T. 8, 14, 15, 16, 21, 41, 48, 49, 53,
    73, 76, 85, &c. II. 4, 22, 58, 60
———— high-coloured, followed by much rain, T. 50, 57, 58, 61, 69.
    II. 23, 68
———— incandescent by electric discharges, II. 307
———— their picturesque appearance in the modification Cumulo-
    stratus, II. 336
Cold, intense or remarkable, T. 44, 94, 115, 117, 119, 120
—— at London and Paris, compared, II. 113
—— greatest degree of, how brought on, II. 117
—— supposed absolute at a certain height, II. 271
—— why greatest just before sunrise, II. 118.
            See Table B

Cold of winter, when established, II. 326
Coke attracts Lightning, T. 48
Column, De Luc's Electrical, Int. xxxi
Comet of 1807, T. 11
Corona Lunar, T. 53, 54, 103, 104, 107, 109, 112, 126. II. 33, 39, 71, 73, 83. II. 343
Cotte, observations by, T. 14, 25, 26, 32
Countries uninhabited, how to find the Temperature of, II. 109
Cuckow, T. 6, 18, 20, 31
———— heard by night, T. 32
Cumming, Alex. his Barometer clock, Int. xiii
Curl-cloud (Cirrus) Int. xxxiii
Currents, in the atmosphere, of the, T. 64. II. 206, 273. See Winds
———————different at different heights, T. 37, 46, 53, 57, 60, 62, 64, 110
—————— their opposition ascertained, T. 61, 67. II. 7
Cyanometer, Int. xxvii. II. 5
Cycle, supposed, of Temperature, II. 289
Cyma, the Cirrostratus compared to a, T. 15, 32, 58. II. 54
Cumulus and Cumulostratus cloud, *passim*
————————————————— defined, II. 330

Dalton, Int. xxvii. II. 177
Daniell, his Hygrometer, II. 178
Darkness, great by smoke, T. 65, 125
Dates, the importance of annexing, to reports, Int. xxxiv
Days, long and short, their relation to Temperature, II. 288
Deer killed by Lightning, T. 83
Declination, Sun's, its effects on Temperature, II. 100, 122—128
—————— Moon's, its effects on the Barometer and Temperature, II. 245—261
————————————————— Rain, II. 249—252
————————————————— Winds, II. 259—270
Depression of the Barometer, great and continued, T. 39
De Luc, Int. xviii
Dew, remarkable, T. 24, 35, 99, 100, 105
———— a guage to measure the, Int. xxvii
———— point, or Vapour-point, Int. xxvii. T. 1, 62, 67. II. 177
———— its refractive power illustrates that of the rainbow, II. 310
Dewy haze, T. 36, 76
Diamonds said to be split by Lightning, T. 93
Diopetous the, probably a meteoric stone, T. 72
Dick, T. L. on an earthquake in Scotland, T. 121
Drains emit an offensive gas before rain, T. 117
Drought, instances of, T. 88, 110, 146
Drowned man figured in the ice, T. 76
Dry air, changes produced by return of, T. 129
———— indicates the approach of frost, II. 170
———— prevents the feeling of intense cold, T. 115

Dry periods, T. 5, 8, 9, 14, 17, 21, 40, 45, 73, 102, 110, 129, 130, 135, 144, 145, 146
—— summer indicated, T. 17
—————— how produced in our climate, II. 305
Ducks, their actions before rain, T. 29
Dust floats in the air before rain, T. 86, 93, 142
Earth's surface, its effect on temperature, II. 131
Earthquake abroad, T. 8, 14, 25, 37, 39, 52, 66, 67, 73, 74, 92, 100, 113, 132, 152, 156
—————— in Scotland, T. 28, 121
—————— on the coast of Devon, T. 79
—————— in Yorkshire, Derbyshire, &c. T. 116
—————— spurious, T. 93, 123, 126
—————— connected with volcanic action, T. 67
Eclipse, solar, its effect on the temperature, T. 125
Ecnephias defined, II. 338
Electrical apparatus, of the, Int. xxix
Electricity, atmospheric, T. 1, 4, 6, 7, 8, 16, 18, 19, 27, 31, 32, 35, 36, 49, 54, 60, 69, 112. II. 24, 49, 56
—————— odour of, perceptible in the air, T. 68, 89, 95, 96, 110, 124
Equinoxes precede the mean temperature, II. 129
—————— on the whole, dry seasons, II. 202
Eruption of Aetna, T. 64
—————— Mount Albay, Philippines, 103
—————— Vesuvius, T. 36, 70
—————— a new volcano in the Azores, T. 19, 53, 58
—————— the Souffrière, St. Vincent's, T. 68
Evaporation, of the, Int. xxiv. II. 162, 171
—————— excessive, T. 40, 44. 63
—————— experiments on, T. 21
—————— gradation of, in a period, T. 136
—————— its effects on the temperature, T. 16. II. 117
—————— and showers characterize the spring, II. 298
Exeter, temperature at, T. 42
Extracts from the Daily papers, how obtained for this work, Int. xxxiv
Extremes of temperature, for the climate, II. 108, 290
————————————— by day and night, II. 119
Experiments with balloons, T. 37, 60, 64
—————— with the insulated kite, II. 24
—————— on the fall of rain, T. 64

Fahrenheit, his scale of temperature preferred, II. 291
Falling Stars. See Meteors.
Fire, how caused by lightning, T. 45
—— in the grate, acceptable at Midsummer, T. 95
—— arms, liable to explode by lightning, T. 45
Flaugergues, his experiments on temperature, II. 282
Floods, T. 6, 22, 27, 28, 52, 59, 78, 110. II. 50
—————— by rain in a higher district, T. 6, 22

Fogs, great or remarkable, T. 89, 138
Forests on fire during heat, T. 61
Forster, Dr. Thos. Int. xxxi. T. 37, 39. II. 19, 24
Franklin, Int. xxxvii
Frost, late or premature, T. 99
―― continued, an exception to the general rule of the climate, II. 292
―― on the ground only, T. 11, 115
―― its elegant drapery, T. 89
Frosty periods, T. 14, 27, 89, 90, 91, 102, 115, 138
―――――― equality of their mean, T. 115
―― nights, their proportion in each month, II. 316—328

Geneva, its local currents, Int. x
Gibson John, T. 11, 39, 122. II. 15
Glory, a meteorological described, II. 345
Gloucester, hailstorm at, T. 21
Gossamer, T. 12, 49, 62, 101, 124
Guages for rain and evaporation, Int. xxi

Hail, great or remarkable, T. 6, 21, 30, 32, 44, 55, 57, 96, 101, 108. II. 6, 17, 19, 50, 209
―― Balls, T. 30. 40
―― curious formation of, observed in a shower, T. 28
―― what kind of, most common in winter, II. 296
―― its approaches prevented by firing gunpowder, T. 6
Halo, Solar, T. 3, 8, 70, 85, 87, 103, 112, 125. II. 57, 77, 79
―― Lunar, *passim*
―― ―――― defined, II. 344
―― ―――― remarkable, T. 42, 105, 112. II. 71, 77, 81
―― ―――― diameter of, measured, T. 53, 55
Hanson, Tho. his observations at Manchester, T. 11—24, 30
Hay-making, observations upon, II. 303
Hay-rick pierced by lightning, T. 93
Haze, peculiar or remarkable, T. 21, 30, 90, 93
―― its connexion with thunder-storms, II. 305
Heat, great or remarkable, T. 21, 61, 96, 132, 144, 145, 146
―― greatest, of the climate, II. 110
―― gradation of, through the year, II. 124
―― whether reflected from the Moon, II. 284
See Table B
Henry, Dr. W. on a storm of wind, &c. T. 141
Herschel's Table: a scheme of the weather so entitled: *see end of Table of Lunations*
Hoar-frost, two kinds of, T. 87
―――――― lies longest on bad conductors, T. 100. II. 81
Honey-dew, how produced, T. 8
Horse-chesnut, double vegetation of the, II. 73
Hot wind, in North America, T. 101
House moved, in situ, by a flood, T. 74
Howard, W. T. 97

Hurricane at sea, T. 61
————— the Bahamas, T. 84
————— Guadaloupe, T. 123
————— St. Lucia, &c. T. 136
————— account of one in England, II. 311—316
Hutton, Dr. his theory of rain why not adopted, II. 207
Hygrometer, of the, Int. xix
————— manner of noting the, Int. vii
————— its monthly averages, &c. II. 173
————— of Daniell, described, II. 178
————— mean of the, for the Climate, II. 294

Ice, evaporation from, II. 169
——— islands of, drifting in the Atlantic, T. 94. II. 51, 55, 59
Iceland, its winds opposed to ours, T. 52
——— weather in, T 133
Ignis fatuus, T. 51. II. 216
Inosculation of clouds with clouds, T. 59, 60, 83, 84, 92, 93, 96, 99, 106, &c.
————— ——— with smoke, T. 88, 103. II. 28, 65
Inundations, T. 28, 57, 91. II. 20

Kamsin, or wind of the desert, T. 88
Kirwan, Int, iv. viii. xxxvii. T. 17
Kite electrical, Int. xxx. II. 24

Lead, blown away from buildings, by the wind, T. 28 ; II. 311
Leeches support a temperature below freezing, T. 126
Leyden phial, how affected by the insulated kite, II. 25
Lightning, accidents by, T. 6, 9, 10, 21, 29, 88. II. 58
————— perceptible at great distances, T. 21, 83
————— singular effects of, T. 27, 29, 34, 47, 58, 93
————— violet coloured, T. 106
Lights in windows, their service in a snow-storm, T. 125
London, its site, II. 288
——— warmer than the adjacent country, II. 91. 288
——— the excess of its heat in each month, II. 103
Low temperature detected by a Thermometer near the ground, T. 11
Lowestoft, observations at, II. 71
Lunar periods, of the, Int. v. II. 221
————— the distribution of rain in, exemplified, T. 133, 134, 135.
See Tables F & G

Manchester, weather at, T. 11, 12, 13, 14, 16, 23, 24
Manheim, Meteorological society of, Int. iv
Marshall, Int. xxxvii
Mean temperature, what, II. 90
——— of the Barometer for London, II. 146
Meteor, *passim*

Meteor, anomalous, T. 50, 53
———— large or remarkable, T. 4, 16, 32, 55, 57, 59, 64, 80, 83, 84, 87, 113, 125. II. 24, 36, 40
———— seen in the day time, II. 43, 83
———— in connexion with earthquake, T. 39, 74, 113
Meteorolites in North America, T. 14
————— near Thoulouse, T. 68
————— in the East Indies, T. 72
Meteorology recommended to men of leisure, Int. xxxvi
Mist, great or remarkable, T. 37, 89. II. 3
Modifications of clouds, remarks on the, II. 331—338
Months, the several, their Meteorological characters, II. 316—328
Moon, the appearances of its disk, as prognostics, T. 28, 36, 41, 47, 50, 54, 70, 82, 84, 91, 110, 112, 113, 126. II. 22, 57, 70
Moon at Full, its light compared with twilight, T. 19, 82
———— reflection from the dark surface of the, T. 81, 130. II. 83, 155, 286
———— its influence on the atmosphere examined, II. 223—270
———— its possible effect on our temperature by radiation, II. 284
Moscow, intense cold at, T. 40
Mushroom large, after thunder and autumnal warmth, T. 123

Nights, calm in fair weather, T. 31
———— frosty, their proportion in each month, II. 316—328
———— warm, indicative of rain, T. 147
———— always warmer in London than around it, II. 121, 289
Nightingale, T. 19, 20, 31
———— sings in a thunder shower, T. 81
Nimbus cloud, *passim*
———— defined, II. 330
———— electricity of, ascertained, T. 6, 19
———— specimens of, T. 25, 30, 39, 56, 111, 118. II. 80
North America, weather in, T. 119
Numerical designation of the months, why used, Int. vi

Observatory, Royal, at Paris, II. 191, 278
Oil Barometer, account of an, Int. xiv

Paraselené, II. 19
Parhelia, T. 67. II. 57, 68, 77. II. 343
Paris, weather at, T. 12, 21, 24, 123
———— its heat and cold compared with ours, T. 21. II. 110, 113
Periods Lunar, why adopted in this work, Int. v. II. 222
————— when discontinued, II. 87
Perspiration, in what condition or the air most obstructed, T. 88
Petersburgh, unusual thaw at, II. 75
Phillips, W. his Lectures, Int. iv
Pictet, M. A. Int. xxvii
————— proposes a congress of Meteorologists, II. 290
Plaistow, observations at, T. 1—76

Plymouth, temperature at, T. 113
Precipice, fall of a, T. 22, 73

Quicksilver frozen in the air, T. 40, 119

Rain, of the, II. 181—220
—— annual mean depth of, at London, II. 183, 294
—— depth of, at London in different years, II. 184, 185
—— connexion of, with the temperature, II. 186
—— —————————— different winds, II. 203
—— estival and autumnal, II. 201
—— excessive, instances of, T. 22, 28, 47, 50, 51, 59, 66, 71, 84,
   97, 101, 120, 122, 126, 143, 148
—— experiments on the fall of, T. 64
—— general over this island, T. 86
—— how produced in these latitudes, II. 205
—— proportion of, for each month, &c. II. 188, 192—198
—— particular tracts affected by, T. 71, 110.
—— volcanoes probably affected by, T. 67
                    See Tables D, H
Rainbow, *passim*
————— different kinds of the, II. 334
————— frequent or unusual exhibition of the, T. 7, 15, 19, 30, 64,
   75, 87, 111
————— manner of the distribution of the light on a, T. 81, 106
————— Lunar, T. 26. II. 35, 68
Rain-guage, of the, Int. xxi
Read, John, Electrician, Int. xxix
Reikiavik, observations at, T. 52
Rime, how produced, Tab. 2. II. 334
—— brings off the leaves in autumn, T. 25
—— elegant appearance of, T. 89
River, changing its course periodically, T. 75
Robertson, Dr. Int. xxxvii
Rod, insulated electrical, of the, Int. xxix
————— its indications. See Electricity
Royal Society, its register, II. 91, 190
Run, remarkable, of a ship in a tempest of wind, II. 315

Sacks of earth, their use in stopping inundation, T. 28
Saussure, Int. xx, xxviii
Saxifraga, different species of, retain the hoar-frost long, T. 100,
   II. 81
Scud, T. 7
Sea, unusual agitation of the, T. 57, 70
Seasons, the four, their duration, mean temperature, &c. II. 129,
   278—284
Shadows of clouds in the air, T. 4, 63, 83, 84. II. 57, 79, 86
Ships struck by lightning, T. 57, 58, 87, 88
Shower of burnt paper, T. 91

Shower, curious composition of a freezing one, T. 28
Showers, their electricity, T. 6, 18, 19, 31, 35. II. 300
———— affect particular situations, T. 71, 110
Site of the observations described, II. 288
Six, James, his Thermometer, Int. xv
Sky coloured, before rain, II. 68
Slip of a bank, T. 22
Smoke, great darkness from, T. 65, 124, 125
———— inosculates with clouds, T. 88, 103. II. 65
Snow, remarkable or great, T. 14, 16, 28, 31, 39, 79, 90, 100, 115, 124
———— cristallised, T. 27, 40, 75, 102. II. 35, 296
———— evaporation of the, T. 16, 28, 52, 90, 115. II. 169
———— introduces dry weather, T. 5, 26, 43
———— its effect on the Temperature, T. 115. II. 42
——— at Messina, uncommon, T. 32
——— observations upon, T. 115
Snow, unmelted, on our mountains, in 1816, T. 120
Snowballs, natural formation of, T. 89
Solar Eclipse, II. 53
———————— lowers the Temperature, T. 125
Solstitial rains, II. 200
Song-birds, few, after a hard winter, T. 94
Soot disappears on snow, in sunshine, T. 126
Spiders, their motion through the air, II. 309
Soufrière, great eruption of the, T. 68
Sounds, unusual propagation of, T. 62, 64, 75
Sponge-like pores in clouds, II. 11
Spots on the Sun's disk, peculiar figure of, T. 107
Spring, its mean temperature, duration, &c. II. 129, 293
———— of 1816 very wet, T. 122
———— at Versailles, II. 7
———— without a frost, T. 20
Stockton, Jas. his observations, at Malton, Yorkshire, T. 30, 44
Storm, dreadful of 1703, II. 311
Storms of wind, T 32, 46, 64, 67, 75, 77, 113, 121, 123, 124, 126, 127, 141. II 3, 36, 37, 40, 46
Stormy weather, T. 27, 28, 36, 39, 57, 78. II. 46, 77, 79
Stratus cloud, *pa sim*
———————— defined, II. 330
———————— before a storm, T. 7, 24
———————— coloured by reflection, T. 63, 78
———————— curious effect of one, T. 110
———————— taken for an inundation, T. 61
Stutgardt, great rain at, II. 219
Summary of the phenomena of the seasons, II. 288
Summer, its duration, mean temperature, &c. II. 130, 301
———— dry, indicated, T. 17
———— wet, of 1816, T. 122
Summer lightning, T. 48, 49

Summers, what winds decide their character, II. 305
Sun's disk distorted by refraction, II. 8
—— rays, their mean effect in each month, II. 123
Sunshine, its effect in giving colour, II. 55
Swallows, T. 6, 12, 18, 25, 31, 43, 47, 49, 56, 62, 74, 105, 110, 111, 118. II. 11, 82
Swithin, T. 83. II. 198

Temperature, II. 89
—————— medium and mean, what, II. 90
—————— mean, of the Climate, II. 91, 288
—————— mean annual, for 20 years, II. 95
—————— mean monthly for London and the environs, II. 99
—————— mean diurnal, its gradation, II. 124
—————— mean, by day and night, in Tables. II. 134—143
—————— its variations in the same season, II. 101
—————— ————— in different periods, II. 144
—————— irregular occurrence of the maximum by day, T. 26, 29, 50, 81, 84, 98, 101, 103, 109, 111, 114, 116, 122, 125. II. 81.   See Tables A, B, G
Terminology of clouds, &c. II. 329
Thames frozen, T. 90
Thermometer, of the, Int. xv
—————— Six's, why preferred, Int. xv
Thomson, Dr. Int. xxxvii
Thunder, before dry weather, T. 8, 39
—————— before wet weather, T. 6, 10, 19, 22, 31, &c.
—————— in winter, T. 4, 39, 42, 66, 91, 102, 114. II. 6
—————— produces a spurious earthquake, T. 93, 123, 126
Thunderstorms, passim
—————— distant, T. 6, 18, 19, 20, 21, 23, &c.
—————— occurring periodically, T. 100
—————— rare in Ireland, T. 58
—————— their magnificence, in summer, II. 306
Tide, high or irregular, T. 2, 12, 57, 114. II. 46
—— Atmospherical, II. 272—276
Tornado, T. 33, 57, 73. II. 20, 50, 340
—————— minute account of a, T. 94
Tottenham, observations at, T. 77—157
Trade winds, of the, II. 273
Trees blown down, T. 32, 101
—— shivered by lightning, T. 21, 59, 64. II. 47
—— shrouded in rime, T. 40, 52
Trombus defined, II. 341
Turbo defined, II. 339
Twilight, phenomena of the, described, T. 83
—————— brilliant or coloured, T. 19, 21, 31, 33, 36, 60—69, 70—87, 100
—————— reflected in the Eastern sky, or clouds, T. 73, 93, 126
—————— with moonlight, their shadows compared, T. 19, 82
Typhon defined, II. 340

Vane, how to construct a good, Int. ix
Vapour, least in quantity at the beginning of winter, **II. 296**
Vapour-point, T. 1, 21, 67, 83. II. 177
———————— how ascertained, Int. xxviii
Variable climate, the advantages of a, II. 292
Variation of the compass, present, Int. xi
Vegetation, notices of the state of, T. 2, 19, 20, 62, 73, 116, 117. II. 83, 84
———————— protracted into the winter, T. 2. II. 73
Vesuvius in action, T. 36, 70, 89
Volcano, a new, T. 19
———————— submarine, T. 53, 58
———————— at Vellas described, T. 68

Warm years dry, and the cold, wet, II. 294
———————— rain before thunder, T. 9
Water newly pumped from a well emits steam, T. 103
Waterspout, T. 46, 58, 71, 109. II. 341
———————— seen from Stratford, II. 15
Wells, Dr. on radiation of heat, II. 117
Wet periods, T. 7, 13, 22, 23, 28, 47, 50, 51, 52, 59, 62, 64, 66, 67, 69, 74, 81, 86, 100, 101, 104, 116, 120, 122, 124, 126, 128, 131, 138, 139, 140, 141, 143, 148
——— summer, how indicated, T. 17
——— spring before a warm summer, II. 301
——— and dry years, II. 186, 305.
See Tab. D, H
Whale on shore in Scotland, T. 27
Whirlwind, T. 57, 120. II. 57, 62, 339
Wild fowl migrating, T. 27, 40
Wind, force of the, T. 28. II. 40
——— modulation of its sound before rain, T. 83, 93, 104, 111, 113, 115
——— veering against the Sun's course, T. 45
Windows, when moist on the outside only, T. 113
Winds, of the, II. 155, 161
——— their yearly and monthly proportions, II. 156, 158
——— in what way subject to the Moon, II. 276
See Tables B, D
Winter, its duration, mean temperature, &c. II. 130, 294
——— late commencement of in 1807, T. 2
——— long duration of in 1819—20, II. 214
——— severe, T. 90, 115, 125
——— ——— in Italy, T 67
——— ——— in North America, T. 119
Wire melted and dispersed by lightning, T. 29, 34, 47

London: printed by W. Phillips,
George Yard, Lombard Street.

Printed in the United States
By Bookmasters

Printed in the United States
By Bookmasters